MOLECULAR BIOLOGY
INTELLIGENCE
UNIT

Abl Family Kinases
in Development and Disease

Anthony J. Koleske, Ph.D.
Department of Molecular Biophysics and Biochemistry
Yale University
New Haven, Connecticut, U.S.A.

LANDES BIOSCIENCE
GEORGETOWN, TEXAS
U.S.A.

SPRINGER SCIENCE+BUSINESS MEDIA
NEW YORK, NEW YORK
U.S.A.

ABL FAMILY KINASES IN DEVELOPMENT AND DISEASE

Molecular Biology Intelligence Unit

Landes Bioscience
Eurekah.com

Copyright ©2006 Landes Bioscience and Springer Science+Business Media, LLC

Please address all inquiries to the Publisher:
Landes Bioscience / Eurekah.com, 810 South Church Street, Georgetown, Texas, U.S.A. 78626
Phone: 512/ 863 7762; Fax: 512/ 863 0081
www.eurekah.com
www.landesbioscience.com

ISBN: 0-387-36640-7

While the authors, editors and publisher believe that drug selection and dosage and the specifications and usage of equipment and devices, as set forth in this book, are in accord with current recommendations and practice at the time of publication, they make no warranty, expressed or implied, with respect to material described in this book. In view of the ongoing research, equipment development, changes in governmental regulations and the rapid accumulation of information relating to the biomedical sciences, the reader is urged to carefully review and evaluate the information provided herein.

Library of Congress Cataloging-in-Publication Data

Abl family kinases in development and disease / [edited by] Anthony J. Koleske.
    p. ; cm. -- (Molecular biology intelligence unit)
Includes bibliographical references and index.
ISBN 0-387-36640-7
1. Protein-tyrosine kinase. 2. Protein-tyrosine kinase--Pathophysiology. I. Koleske, Anthony J. II. Series: Molecular biology intelligence unit (Unnumbered)
[DNLM: 1. src-Family Kinases--metabolism. 2. Cell Physiology--drug effects.
3. src-Family Kinases--physiology. QU 141 A148 2006]
QP606.P78A25 2006
612'.01516--dc22

                  2006020098

Dedication

To Ruth
and
Emily and Benjamin

CONTENTS

EDITOR

Anthony J. Koleske, Ph.D.
Department of Molecular Biophysics and Biochemistry
Yale University
New Haven, Connecticut, U.S.A.
Email: Anthony.Koleske@yale.edu
Chapter 5

CONTRIBUTORS

Shahin Emami
INSERM
U482
Saint-Antoine Hospital
Paris, France
Chapter 6

Oliver Hantschel
Center for Molecular Medicine
    of the Austrian Academy of Sciences
Vienna, Austria
Email: ohantschel@cemm.oeaw.ac.at
Chapter 1

Richard L. Klemke
Department of Immunology
Scripps Research Institute
La Jolla, California, U.S.A.
Email: klemke@scripps.edu
Chapter 6

Yosuke Minami
Division of Biological Sciences
University of California-San Diego
La Jolla, California, U.S.A.
Chapter 4

Eva Maria Y. Moresco
Department of Molecular Biophysics
    and Biochemistry
Yale University
New Haven, Connecticut, U.S.A.
Chapter 8

Ann Marie Pendergast
Department of Pharmacology
    and Cancer Biology
Duke University Medical Center
Durham, North Carolina, U.S.A.
Email: pende014@mc.duke.edu
Chapter 2

Ruibao Ren
Department of Biology
Rosenstiel Basic Medical Sciences
    Research Center
Brandeis University
Waltham, Massachusetts, U.S.A.
Email: ren@brandeis.edu
Chapter 7

Martin Alexander Schwartz
Department of Microbiology
University of Virginia
Charlottesville, Virginia, U.S.A.
Email: maschwartz@virginia.edu
Chapter 3

Giulio Superti-Furga
Center for Molecular Medicine
    of the Austrian Academy of Sciences
Vienna, Austria
Email: gsuperti@cemm.oeaw.ac.at
Chapter 1

Keith Quincy Tanis
Department of Psychiatry
Yale University
New Haven, Connecticut, U.S.A.
Email: Keith.Tanis@yale.edu
Chapter 3

Cheryl L. Thompson
Department of Cell Biology
Program in Neuroscience
Harvard Center for Neurodegeneration
  and Repair
Harvard Medical School
Boston, Massachusetts, U.S.A.
Chapter 9

David Van Vactor
Department of Cell Biology
Harvard Medical School
Boston, Massachusetts, U.S.A.
Email: david_vanvactor@hms.harvard.edu
Chapter 9

Jean Y.J. Wang
Division of Hematology-Oncology
Department of Medicine
University of California-San Diego
La Jolla, California, U.S.A.
Email: jywang@ucsd.edu
Chapter 4

Jiangyu Zhu
Division of Biological Sciences
University of California-San Diego
La Jolla, California, U.S.A.
Chapter 4

PREFACE

Abl family nonreceptor tyrosine kinases serve as essential relays between the world outside of the cell and the cellular machinery that coordinates changes in cytoskeletal structure, regulates cell division, responds to DNA damage, and executes programmed cell death. Mutational activation of these kinases subverts normal cellular controls on these processes and leads to human leukemia. The book is a compilation of what we presently know about the cellular and molecular mechanisms by which Abl family kinases function in development and how dysregulation of these processes causes leukemia in humans and developmental defects in animals.

Protein kinases transmit information by phosphorylating specific substrates in response to discrete stimuli. The activity of Abl family kinases is normally kept under very tight control in cells, but can be induced by diverse stimuli, including growth factor or adhesion receptor engagement, DNA damage, and oxidative stress. In Chapter 1, Oliver Hantschel and Giulio Superti-Furga begin our book with a scholarly discussion of how Abl family kinase activity is regulated by intra- and intermolecular interactions. This theme is continued in Chapter 2 by Ann Marie Pendergast as she summarizes pioneering work from her lab examining how Abl family kinases become activated by growth factor receptors. Keith Tanis and Martin Schwartz complement this discussion by reviewing how Abl family kinases mediate signaling from integrin adhesion receptors in Chapter 3.

Upon activation by various stimuli, Abl family kinases interact with downstream effector pathways to execute diverse cellular functions. Two principal functions of Abl family kinases are the regulation of programmed cell death and the control of cytoskeletal dynamics. In Chapter 4, Jean Wang, Yosuke Minami, and Jiangyu Zhu rigorously explore work from their lab and others detailing with the signaling pathways by which Abl regulates cell death. I review examples where Abl family kinases have been shown to regulate cell morphogenesis and review the biochemical mechanisms by which Abl and Arg control F-actin rearrangements in Chapter 5. Shahin Emami and Rich Klemke in Chaper 6 enrich this discussion by reviewing how Abl family kinases coordinate these cytoskeletal pathways to control directed cell migration.

Oncogenic activation of Abl family kinases causes leukemia in vertebrates. Ruibao Ren in Chapter 7 provides a comprehensive discussion of oncogenic variants of Abl family kinases, the cell types they target, and the mechanisms by which they lead to cancer.

Genetic studies have shown that Abl family kinases are important regulators of cell development and morphogenesis, particularly in the hematopoietic and nervous systems. In Chapter 8, Eva Moresco examines the genetic studies that implicate Abl family kinases in diverse aspects of vertebrate development. Cheryl Thompson and David Van Vactor review the pathways and mechanisms by which Abl family kinases control axon and dendrite morphogenesis in Chapter 9.

This book will provide the reader with an exciting entry into the cellular functions of Abl family kinases and the developmental and disease processes that depend on their function. Students who read this book will gain exposure to an uniquely interdisciplinary field that engages structural biologists, chemists, cell and molecular biologists, medical oncologists, geneticists, and neuroscientists. Each author has also offered his or her unique perspective on future challenges in their discipline. The good news for students is that despite everything we have learned about Abl family kinases in the past 30+ years, there are many fundamentally important problems that remain to be solved.

Anthony J. Koleske
New Haven, Connecticut, U.S.A.

Acknowledgments

I wish to thank Elizabeth Vellali for expert research and administrative support and members of my lab, past and present, for stimulating discussions.

CHAPTER 1

Mechanisms of Activation of Abl Family Kinases

Oliver Hantschel* and Giulio Superti-Furga

Abstract

E
vidence that has accumulated over the last years points to c-Abl and Arg (ABL1 and
ABL2) as being particular forms of the Src family of kinases. Just as much as or even
more than the Src kinases, Abl members are built to be able to couple protein-protein
interaction with protein tyrosine kinase catalytic output. This stems from the constant competition between self-inhibitory *intra*molecular interactions (mostly via the SH3 and SH2 domains) and generally activating *inter*molecular interactions with ligands. Ligand engagement
both regulates and, in turn, is regulated by the level of activity of the kinase domain. A series of
post-translational modifications act on this balance and allows the integration of catalytic activity, localization and multiprotein complex assembly functions. Most excitingly, the majority
of the principles appearing to govern c-Abl and Arg are still operational in the Bcr-Abl oncogenic counterpart and affect the efficacy of small molecular ATP-competitors.

The Abl family of tyrosine kinases is regulated by a complex set of intramolecular interactions
that impinge both directly and indirectly on the Abl kinase domain and lead to effective inhibition of tyrosine kinase activity both in vitro and in vivo. Even a partial, albeit persistent, disruption of these autoinhibitory constraints results in cell transformation and different forms of cancer in humans. The fusion-proteins Bcr-Abl, Tel-Abl and v-Abl are three well-characterized examples
in this respect. Here, the kinase activity is mostly switched on, contributing to the deregulation of
cell growth. On the other hand, the controlled activation of Abl kinases is required for a large
number of normal cellular processes. The most important ones are of central interest to many
research groups and are discussed extensively in other chapters of this book. In this chapter, we
provide an overview of the mechanisms by which multiple cellular proteins transiently activate
Abl kinases to perform cellular functions. We present the entire set of mechanisms that lead to
Abl activation, grouping the numerous studies on physiological stimuli acting on Abl into distinct activation categories. The recently obtained insights into the structure of autoinhibited Abl
is integrated and is used as guide to explain the different molecular mechanisms.

Structure and Regulation of c-Abl

The state of low catalytic tyrosine kinase activity of c-Abl and Arg is the result of particular
restraints on the kinase domain imparted by other regions of the protein. This is exemplified by
the fact that the kinase domain alone has a 10-100-fold higher catalytic activity than the
full-length protein (Superti-Furga laboratory, unpublished observation). These additional regions
of the protein mediate both intra-, as well as intermolecular interactions that either directly or

*Corresponding Author: Oliver Hantschel—Center for Molecular Medicine of the Austrian
Academy of Sciences Lazarettgasse 19/3 1090 Vienna, Austria.
Email: ohantschel@cemm.oeaw.ac.at

Abl Family Kinases in Development and Disease, edited by Anthony Koleske.
©2006 Landes Bioscience and Springer Science+Business Media.

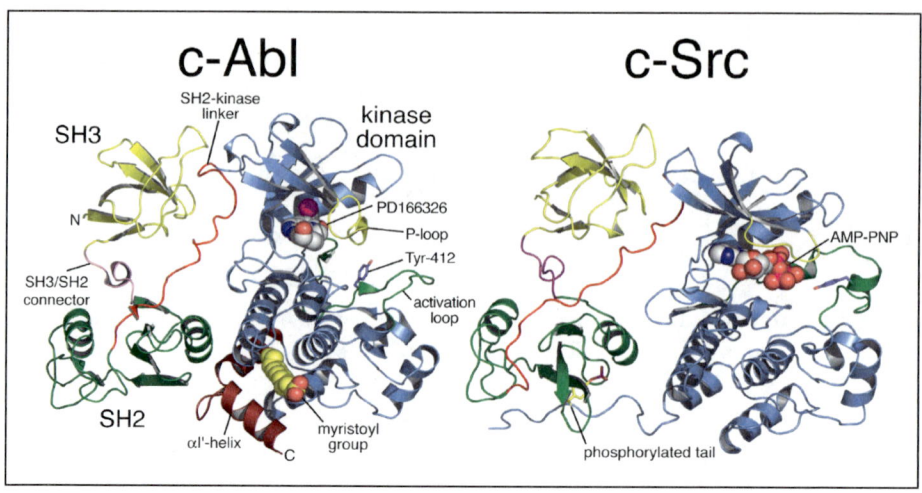

Figure 1. Structures of regulated c-Abl and c-Src. Ribbon representation of regulated c-Abl in complex with the kinase inhibitor PD166326 (left; PDB entry 1OPK,[1]) compared to regulated c-Src in complex with the ATP analogue AMP-PNP (right; PDB entry 2SRC,[68]). The SH3, SH2 and kinase domains are shown in yellow, green and blue, respectively. The SH2–kinase domain linker is shown in red, the SH3–SH2 connector in purple and the carboxy-terminal helices αI and αI' of c-Abl in dark red. The P-loop and the activation loop are yellow and green, respectively. In c-Src, the SH2 domain binds the tyrosine-phosphorylated tail emerging from the kinase domain, whereas in c-Abl, the SH2 domain is closely docked to the kinase domain. The myristoyl group of c-Abl binds to a deep hydrophobic pocket in the kinase domain and induces a conformational switch (dark red helices) that allows docking of the SH2 domain. The structure figures were prepared using PyMol.[69]

indirectly impose a restriction on the kinase domain. A major contribution to autoinhibition of c-Abl is provided by a series of intramolecular interactions of the SH3 and SH2 domains of c-Abl, which bind to the distal face of the kinase domain, opposite of the substrate- and ATP-binding sites, and act as a "clamp" that keeps the kinase domain in a conformation of low catalytic activity.[1-3] This domain arrangement in c-Abl closely resembles that seen in the structures of regulated Src-family kinases[4-6] (Fig. 1).

The SH3 domain binds the linker segment that connects the SH2 and the kinase domain (the SH2–kinase linker, which adopts a polyproline type II helix) and the small lobe of the kinase domain.[7] The SH2 domain, on the other hand, forms a tight protein–protein interface with the large lobe of the kinase domain.[1] As observed in structures of the isolated kinase domain, the kinase domain alone would not be compatible with SH2-domain docking, as the C-terminal helix αI clashes with the SH2 domain.[1] A crystal structure of the isolated kinase domain in complex with a myristoylated peptide corresponding to the very amino terminus of c-Abl showed that binding of the myristoyl group to a deep hydrophobic pocket in the C-lobe can break the αI-helix and form another helix, αI'. This helix has a very different orientation and thereby renders the docking site accessible for the SH2 domain[8] (Fig. 1). In turn, binding of the myristoyl group to the pocket seems to be absolutely essential for adopting an assembled, autoinhibited conformation of c-Abl 1b, and forms of c-Abl 1b that lack the myristoyl group therefore show constitutive tyrosine kinase activity.[8]

Besides their role as intramolecular regulators of kinase activity, both the SH3 and SH2 domains have been shown to interact with a large number of other proteins (reviewed in ref. 9). Binding of the SH3 and SH2 domains to their respective ligands in these proteins and thereby disrupting the intramolecular Abl interactions appear to be a widely used mechanism of Abl activation.

Activation of Abl by SH3 Domain Ligands

The Abl SH3 domain participates in a sandwich interaction, binding the SH2-kinase linker, which adopts a polyproline type II helix, between itself and the amino-terminal lobe of the kinase domain.[7,1] In addition to the data from the crystal structures and molecular dynamics simulation,[1] mutagenesis of the SH3-domain, the SH2-kinase linker or the small (amino-terminal) lobe of the kinase domain show that perturbance on any participant in this intramolecular interaction leads to activation of Abl kinase- and transformation activity.[7,8,10,11] As for Src-family kinases, the kinase domain of Abl needs to be able to adopt the proper conformation in order for the linker and SH3 domain to "dock". In c-Src and in Lck, the kinase domain stabilized in a state of high-activity by, for example, phosphorylation of the activation loop is less prone to support SH3/linker docking.[12-14] This three-way interaction ensures high specificity to the "sandwich". Interestingly, recent evidence suggests that the SH3-linker-kinase interaction is at least partly maintained in the otherwise highly de-regulated Bcr-Abl.[15]

In order to disrupt the intramolecular interaction of the SH3-domain with the linker, competition with a proline-rich ligand that is provided in trans, i.e., by another protein, can result in activation. Many examples for proteins that contain SH3-domain ligands and are able to activate c-Abl have been studied in detail, including c-Cbl, c-Jun, Dok-R, RFX1 and ST5.[16-20] Unfortunately, many of these studies fail to show that upon mutation of the PxxP motif Abl activation is impaired. In addition, activation in vitro with an SH3 domain ligand peptide has not yet been reported, probably due to the low level of affinity of SH3 peptide ligands for their respective binding site. Most SH3 domain-dependent activators are, in turn, also phosphorylated by c-Abl. These phosphorylated residues could bind to the Abl SH2 domain, thereby further activating c-Abl kinase activity.

Activation of Abl by SH2 Domain Ligands

SH2 domains bind tyrosine-phosphorylated peptides in a particular sequence specific context, as these domains contain two surface pockets, one that recognizes the phospho-tyrosine and a second one that binds to specific residues that are located downstream of the phospho-tyrosine.[21,22]

The SH3 domain of Abl is already bound to its respective ligand in autoinhibited Abl. In contrast, the SH2 domain has no ligand engaging the phospho-tyrosine-binding pocket in autoinhibited c-Abl. This is due to the tight protein-protein interface of the SH2 domain with the carboxy-terminal lobe of the kinase domain, which is stabilized by a network of interlocking hydrogen bonds and partly occludes access to the SH2 domain, as the kinase domain masks the binding sites for the residues that are immediately upstream of the phospho-tyrosine.[1,8] Based on the Abl crystal structures, binding of phosphopeptides to the SH2 domain in autoinhibited c-Abl was predicted to disrupt the SH2-kinase domain interface and possibly result in kinase activation.[1]

In support of this, and similar to the situation for c-Src, tyrosine-phosphorylated peptides that were derived from c-Abl substrates are capable of activating c-Abl in vitro. A concentration-dependent increase in c-Abl activity was observed and required an intact SH2-domain.[8] This unexpected finding offered an explanation for the repeatedly observed activation of c-Abl by high concentration of substrates. Examples found in the literature include Abl interactor 1 (Abi-1), Nck, paxillin, Cbl, γ-PAK, Cables and possibly also p73, c-Crk, Ena (enabled), and disabled.[16,23-34] Furthermore, this may form the basis for the described nuclear tyrosine phosphorylation circuit involving c-Abl, its substrate c-Jun and the MAP-kinase JNK.[17] SH2 domain-dependent interactions of c-Abl with activated EphB2 and Trk receptor tyrosine kinases,[35,36] as well as the B-cell receptor ancillary protein CD19[37] were also described and are likely to reflect the activation mechanism of c-Abl by receptor tyrosine kinases. However, for most of the examples presented above, the phosphorylation site in the substrates/activators has not been mapped and confirmed by mutating the site. Therefore, the precise molecular mechanism awaits further analysis.

This ability of SH3 ligands (low affinity) and SH2 ligands (high affinity only after phoshorylation) to activate c-Abl raises many important questions. What is the typical order of events? What is the trigger and what the "stabilizing" event? Mechanistically, substrates of c-Abl may initially be phosphorylated through the basal kinase activity of c-Abl or other kinases and initiate a positive feedback loop, which involves SH2 domain-dependent activation of c-Abl kinase activity and results in facilitated substrate recruitment, thereby shifting c-Abl into an active conformation and preventing it from reassuming the regulated, inhibited conformation (Fig. 2). In line with this, efficient phosphorylation by c-Abl seems to require a functional SH2 domain[10,38] and forms of c-Abl with a SH2 domain mutation display reduced in vitro kinase activity (Superti Furga laboratory, unpublished observation).

Similar to its role in kinases of the Src-family, the Abl SH2 domain also appears to be important for processive phosphorylation as substrates that contain multiple phosphorylation sites are more efficiently phosphorylated than substrates with only single phosphorylation sites.[39-41] Although differently interpreted in the past, a "priming" site may initially be phosphorylated through the basal kinase activity of c-Abl or other kinases and initiate the positive feedback loop described above (Fig. 2). The substrate specificity of nonreceptor tyrosine kinases and the binding specificity of their associated SH2 domain strongly correlate.[42,43] This means that good Abl substrates are also good binders to the Abl SH2 domain and vice versa. This supports the known concept of coevolution of SH2-binding and kinase specificities.[43] However, it is important to note though that most Abl phosphorylation sites that are found in physiological substrates do not match the optimal substrate sequence.

Interestingly, most of the proteins that were shown to activate c-Abl by binding of a PxxP motif to the Abl SH3 domain also become phosphorylated by Abl and subsequently serve as SH2 domain-dependent activators. Vice versa, most proteins that act as SH2 domain-dependent activators also contain SH3 domain ligands. Is simultaneous SH3 and SH2 binding an obligatory mechanism for most interactors? It will be interesting to test the relevance of the proper "spacing" between the phosphotyrosine and the SH3 binding determinant. Is there a typical distance? Does engagement by the SH2 domain and SH3 domain necessarily occur in cis on the same substrate/activator molecule? The outcome is in all cases a release of the SH3-SH2 domain clamp from the kinase domain resulting in efficient and localized activation of c-Abl. Although this dual role was shown only for a few cases yet, we hypothesize that it represents the general mechanism of how Abl substrates/activators act. Therefore, both activation parameters seem to be intimately connected and in most cases the order of events is indistinguishable under the current low resolution level of our analytical tools.

Activation of Abl by Phosphorylation

One of the major differences between c-Abl and its oncogenic counterparts is that c-Abl is normally not phosphorylated on tyrosine residues. The activation loop of the Abl kinase domain folds into the active site and prevents binding of both the substrate and ATP.[44] In most kinases, phosphorylation of one or more residues in the activation loop stabilizes it in a conformation that serves as a binding platform for the peptide substrate and facilitates the phosphotransfer reaction.[45] In c-Abl, phosphorylation of Tyr412 in the activation loop is coupled to a concomitant increase in catalytic activity and can occur by an autocatalytic mechanism in trans.[46,47] Furthermore, treatment of cells with platelet-derived growth factor (PDGF) results in activation of c-Abl through measurable Tyr412 phosphorylation by Src-family kinases.[46,48,49] Besides Tyr412, the second proline of the PXXP motif of the SH2–kinase linker to which the c-Abl SH3 domain binds, is replaced with a tyrosine (Tyr245) that can be phosphorylated. Mutation of Tyr245 inhibited the autophosphorylation-induced activation of wild-type c-Abl by 50%.[47,50]

In addition to phosphorylation on Tyr412 and Tyr245, active forms of c-Abl, such as c-Abl that is activated by constitutively active Src, as well as Bcr-Abl and v-Abl are phosphorylated on more than 20 residues (reviewed in refs. 3,51-53 and Superti-Furga laboratory, unpublished results). Interestingly, besides tyrosine residues, many serine and threonine residues are

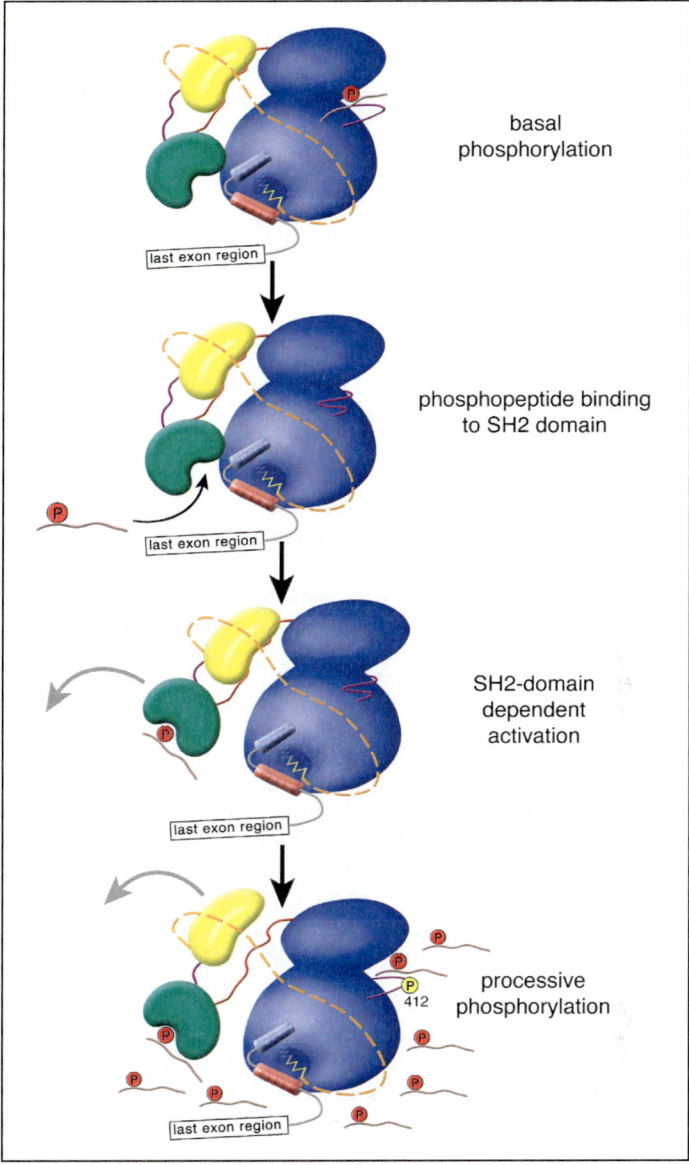

Figure 2. SH2-domain dependent positive feed-back loop on catalytic activity of c-Abl. From top to bottom, the cartoon shows a possible order of event in the activation of c-Abl by SH2 domain engagement. An initial phosphorylation event may arise through basal activity of the enzyme and accessibility of a particular substrate to the Abl kinase domain. Any of the positive events listed in Figure 3 may apply as well, as, for example, engagement of the SH3 domain by a binder competing with the intramolecular interactions. Once, phosphorylated, a substrate may engage the SH2 domain of the Abl kinase, leading to a higher and more stable level of activity. This in turn, may result in phosphorylation of additional sites within the same substrate molecule or of adjacent proteins. These sites may represent even better or just additional binding partners for the same or other Abl proteins. SH2-domain dependent "processive" phosphorylation has been described previously in vitro as well as in vivo.[39-41] Engagement of a tyrosine phosphatase may efficiently oppose the process.

also phosphorylated. For the vast majority of these phosphorylation events, the responsible kinase is unknown. Moreover, the functional consequences of these phosphorylation events on activity, structure and regulation of c-Abl have not yet been studied systematically. In particular, whether a possible phosphorylation event would be able to initiate activation by itself or whether it would merely stabilize an activated conformation of c-Abl is not understood. Mapping of the phosphorylation sites on the structure of autoinhibited c-Abl shows that many of these sites are not easily accessible by a cognate kinase and conformational changes are required to accommodate the phosphate group. This suggests that most phosphorylation events in c-Abl might rather stabilize a conformation associated with activity, rather than being able to initiate activation per se.

Once c-Abl is activated, its kinase activity can be downregulated by different mechanisms. Recruitment of PEST-type phospho-tyrosine phosphatases via the SH3-domain containing adaptor protein PSTPIP1 leads to dephosphorylation and subsequent downregulation of Abl kinase activity, presumably by acting on Tyr 412 in the activation loop.[54] In addition, activated forms of c-Abl are rapidly degraded by ubiquitin-dependent proteasomal degradation, whereas the wild-type protein is more stable.[55] Indeed, inhibition of the 26S proteasome lead to an increased level of c-Abl protein.

Activation by Oligomerization

Activation of kinases by proximity-induced trans-phosphorylation has emerged as the most important mechanism in the activation of receptor tyrosine kinases.[56] In the oncogenic fusion protein Bcr-Abl, an N-terminal coiled-coil domain causes tetramerization of the protein, which leads to an activation of Abl kinase activity.[57-59] Other oncogenic Abl fusion proteins, such as Tel-Abl and v-Abl are multimerizing via the PNT (*pointed*) domain and the Gag domain, respectively. Moreover, direct fusion of the Bcr coiled-coil domain or other dimerization motifs to c-Abl leads to activation of kinase activity.[57,60,61] Conversely, deletion of the coiled-coil domain reduces kinase activity, transformation and oncogenicity of Bcr-Abl.[15,57,62]

Regulation of Bcr-Abl

Several different point mutations in the Abl kinase domain lead to clinical resistance to STI-571/Gleevec in CML patients expressing Bcr-Abl.[63] Many of these point mutations interfere with the drug-binding directly by sterically blocking the drug binding site. On the other hand, resistance mutations that are more distal to the drug-binding site seem to lead to resistance by allosteric mechanisms, stabilizing an active conformation of the Abl kinase domain that is not recognized by Gleevec. A systematic investigation of Bcr-Abl resistance mutations provided considerable insight into the mechanisms that underlie the regulation and potential structural domain arrangements in Bcr-Abl. 112 distinct amino-acid substitutions that lead to STI-571-resistant forms of Bcr-Abl have been identified in an unbiased genetic screen in cultured cells.[64] Recovery of almost the complete set of clinically-observed STI-571-resistance mutations validated the approach. Probably one of the most exciting turns in recent times has been the realization that residues conferring STI-571 resistance are identical to residues that deregulate c-Abl kinase activity.[7,8,11,65,66] This strongly implies that mutation of residues that are crucial for c-Abl autoinhibition is capable of rendering Bcr-Abl resistant to STI-571, and indicates that mechanisms governing Bcr-Abl regulation might be identical to those observed in c-Abl. Bcr-Abl can thus be viewed as a form of c-Abl for which the equilibrium is shifted far, but not fully, towards the very high end of the spectrum of activity. This interpretation is also supported by the observation that point mutations in either the SH3 ligand-binding site or the SH2-kinase-domain-linker could restore kinase activity, transformation and leukaemogenesis to dimerization-defective Bcr–Abl.[15]

Conclusions

From everything we have learned so far, it appears that the Abl family of tyrosine kinases are not merely regulated by dramatic on/off switches and the view of an autoinhibited and an active conformation appears to be an oversimplification. Rather, the proteins are capable of

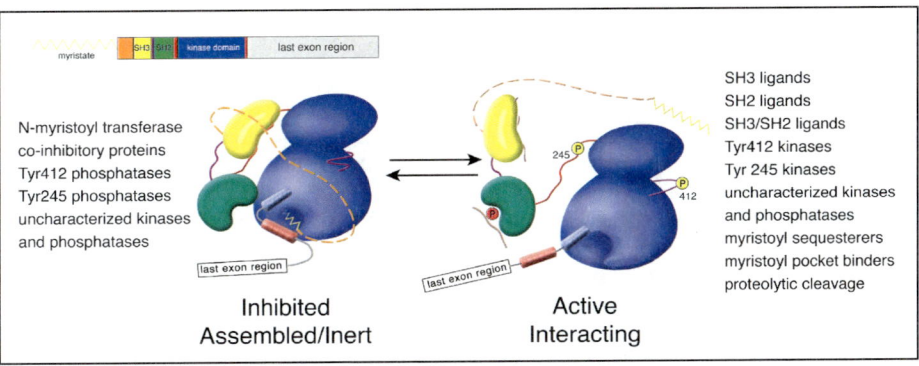

Figure 3. Abl kinases at equilibrium between states of high and low activity, with forces acting on the equilibrium. We envisage c-Abl at constant equilibrium between a self-constrained, essentially auto-inhibited conformation associated with low catalytic activity (left) and a much less constrained conformation where most of the intra-molecular interactions are replaced with interaction with external ligands and the kinase domain has a higher catalytic output. Listed on the left and right side of the cartoon are all those events that we currently believe may pull and push the equilibrium towards the one or the other state of activity. Of the "forces" likely to shift c-Abl into the more inert and inactive conformation much less is known than in the activating events. The PTP-PEST phosphatase is not listed separately as there are reasons to believe to be a prototypic Tyr 412 phosphatase.[70] We know fairly little about the protein kinases and phosphatases governing the phosphorylation state of the 20 or so sites found to be phosphorylated in various forms of Abl (see refs. 3,51-53). However, it is plausible that some of these sites will affect the activity of the enzyme in a negative way. How the various individual proteins reported to affect Abl activity negatively may act, remains to be determined. At least in principle, some of these proteins may act on the activity equilibrium of the cartoon by, for instance, "bridging" the catalytic and regulatory domains thus stabilizing the inactive conformation.[67] The N-myristoyl-transferase is mentioned in the list, as the c-Abl 1b form requires the myristate at its N-terminus to adopt the inhibited conformation.[8] We know more on the events and interactions that are believed to act positively on the activity of Abl kinases (list on the right, see main text for details). Very speculative is the suggestion of the existence of proteins able to bind the myristate moiety at the N-terminus of the c-Abl protein, and "sequestering" it from the internal pocket, leading to activation. Just as unproven is the existence of possible proteolytic events, relieving c-Abl from the constraints mediated by the "cap" region or the SH2-catalytic domain linker. Such events would result in activation of c-Abl. The most straightforward mechanisms of activation, however, appear to either involve phosphorylation or engagement of the SH3 and/or SH2 domains. Several considerations, including ones on accessibility, suggest that phosphorylation of Tyr 412 should be the most direct mechanism of activation, while phosphorylation of Tyr 245 in the SH2-catalytic domain linker may rather act by stabilizing an already "primed" active configuration.

integrating different types and degrees of protein-protein interaction as well as different types and degrees of post-translational modification into a continuum of activity levels, from "very low" (fully inhibited) to "very high" (fully activated) (Fig. 3). Activating and inhibiting proteins that bind to the various domains and sequence motifs of c-Abl, as well as kinases and phosphatases that alter its phosphorylation status may tip the balance further in one or the other direction (Fig. 3). If, as we assume, the kinase domain can adopt a wide series of intermediate conformations resulting in different activity levels, then most regulatory events can be envisaged as forces that shift the equilibrium between fully inhibited and fully activated. The large set of intramolecular interactions that have been described in the last several years seem to have evolved for the precise purpose of coupling environmental cues to activity. Rather harsh interferences with this well-balanced system, like the loss of regulatory motifs and domains, or changes in the oligomerization state of the protein by fusion events, as in the oncogenic v-Abl, Tel-Abl and Bcr-Abl, are necessary to shift the equilibrium towards a more active conformation. On the other end of the spectrum, small-molecule inhibitors like STI-571/Gleevec trap c-Abl in an inactive conformation with the activation loop not phosphorylated and the intramolecular regulatory machinery assembled as shown by the crystal structures.

Future Perspectives

Clearly, among the features that require future attention are the quantitative aspects of regulation and activation of Abl family kinases. What is their actual concentration in the different cellular compartments? What is the stoichiometry of the different interactors? What is the stoichiometry of the post-translational modifications? What are the kinetic parameters for the distinct forms of Abl in the different configurations? Of all parameters affecting the activity of Abl kinases, which are the ones having the strongest effect? Is there a hierarchy in these effects (which are dominant, additive, synergistic, or mutually exclusive)? Another significant and hitherto relatively neglected aspect concerns the temporal and spatial aspects of these activation mechanisms. It will also be important to study at the structural-molecular level how proteins reported to contribute to the inhibited conformation of c-Abl may act.[67] Moreover, there are reasons to suspect, from anecdotal unexplained experimental findings, that yet other intramolecular interactions occur within the entire c-Abl molecule that await further investigation. Finally, it will be very important to obtain three-dimensional insight into the conformation that c-Abl adopts in its activated state.

Judging from the very exciting last few years, it is likely that studies on c-Abl activation mechanisms will remain at center stage of signal transduction and molecular oncology research. From a better understanding of the inhibition and activation mechanisms of c-Abl and its highly active counterpart Bcr-Abl, we can expect both further insights into the molecular mechanism-of-action of anti-Bcr-Abl small molecule inhibitors and possibly even suggestions for improved therapeutic strategies.

Acknowledgements

We thank Lily Remsing-Rix, Tilmann Buerckstuemmer and Alberto Calabro' for critical reading of the manuscript and all present and past members of the Superti-Furga laboratory for stimulating discussions.

References

1. Nagar B, Hantschel O, Young MA et al. Structural basis for the autoinhibition of c-Abl tyrosine kinase. Cell 2003; 112(6):859-71.
2. Harrison SC. Variation on an Src-like Theme. Cell 2003; 112(6):737-40.
3. Hantschel O, Superti-Furga G. Regulation of the c-Abl and Bcr-Abl Tyrosine Kinases. Nat Rev Mol Cell Biol 2004; 5(1):33-44.
4. Xu W, Harrison SC, Eck MJ. Three-dimensional structure of the tyrosine kinase c-Src. Nature 1997; 385:595-601.
5. Sicheri F, Moarefi I, Kuriyan J. Crystal structure of the Src family tyrosine kinase Hck. Nature 1997; 385(6617):602-9.
6. Williams JC, Weijland A, Gonfloni S et al. The 2.35 Å crystal structure of the inactivated form of chicken Src: A dynamic molecule with multiple regulatory interactions. J Mol Biol 1997; 274(5):757-75.
7. Barilá D, Superti-Furga G. An intramolecular SH3-domain interaction regulates c-Abl activity. Nature Genet 1998; 18:280-282.
8. Hantschel O, Nagar B, Guettler S et al. A Myristoyl/Phosphotyrosine switch regulates c-Abl. Cell 2003; 112(6):845-57.
9. Pendergast AM. The Abl family kinases: Mechanisms of regulation and signaling. Adv Cancer Res 2002; 85:51-100.
10. Mayer BJ, Baltimore D. Mutagenic analysis of the roles of SH2 and SH3 domains in regulation of the Abl tyrosine kinase. Mol and Cell Biol 1994; 14:2883-2894.
11. Pluk H, Dorey K, Superti-Furga G. Autoinhibition of c-Abl. Cell 2002; 108(2):247-59.
12. Gonfloni S, Williams JC, Hattula K et al. The role of the linker between the SH2 domain and catalytic domain in the regulation and function of Src. EMBO J 1997; 16:7261-7271.
13. Gonfloni S, Frischknecht F, Way M et al. Leucine 255 of Src couples intramolecular interactions to inhibition of catalysis. Nat Struct Biol 1999; 6(8):760-4.

14. Gonfloni S, Weijland A, Kretzschmar J et al. Crosstalk between the catalytic and regulatory domains allows bidirectional regulation of Src. Nat Struct Biol 2000; 7(4):281-286.
15. Smith KM, Yacobi R, Van Etten RA. Autoinhibition of Bcr-Abl through Its SH3 Domain. Mol Cell 2003; 12(1):27-37.
16. Miyoshi-Akiyama T, Aleman LM, Smith JM et al. Regulation of Cbl phosphorylation by the Abl tyrosine kinase and the Nck SH2/SH3 adaptor. Oncogene 2001; 20(30):4058-69.
17. Barilá D, Mangano R, Gonfloni S et al. A nuclear tyrosine phosphorylation circuit: c-Jun as an activator and substrate of c-Abl and JNK. EMBO J 2000; 19(2):273-81.
18. Master Z, Tran J, Bishnoi A et al. Dok-R binds c-Abl and regulates Abl kinase activity and mediates cytoskeletal reorganization. J Biol Chem 2003; 278(32):30170-9.
19. Katan Y, Agami R, Shaul Y. The transcriptional activation and repression domains of RFX1, a context-dependent regulator, can mutually neutralize their activities. Nucleic Acids Res 1997; 25(18):3621-8.
20. Majidi M, Hubbs AE, Lichy JH. Activation of extracellular signal-regulated kinase 2 by a novel Abl-binding protein, ST5. J Biol Chem 1998; 273(26):16608-14.
21. Waksman G, Shoelson SE, Pant N et al. Binding of a high affinity phosphotyrosyl peptide to the Src SH2 domain: Crystal structures of the complexed and peptide-free forms. Cell 1993; 72(5):779-90.
22. Kuriyan J, Cowburn D. Modular peptide recognition domains in eukaryotic signaling. Annu Rev Biophys Biomol Struct 1997; 26:259-88.
23. Juang JL, Hoffmann FM. Drosophila abelson interacting protein (dAbi) is a positive regulator of abelson tyrosine kinase activity. Oncogene 1999; 18(37):5138-47.
24. Smith JM, Katz S, Mayer BJ. Activation of the abl tyrosine kinase in vivo by src homology 3 domains from the src homology 2/Src homology 3 adaptor Nck J Biol Chem 1999; 274(39):27956-62.
25. Lewis JM, Schwartz MA. Integrins regulate the association and phosphorylation of paxillin by c-Abl. J Biol Chem 1998; 273(23):14225-30.
26. Roig J, Tuazon PT, Zipfel PA et al. Functional interaction between c-Abl and the p21-activated protein kinase gamma -PAK. Proc Natl Acad Sci USA 2000; 97(26):14346-51.
27. Zukerberg LR, Patrick GN, Nikolic M et al. Cables links Cdk5 and c-Abl and facilitates Cdk5 tyrosine phosphorylation, kinase upregulation, and neurite outgrowth. Neuron 2000; 26(3):633-46.
28. Yuan ZM, Shioya H, Ishiko T et al. p73 is regulated by tyrosine kinase c-Abl in the apoptotic response to DNA damage. Nature 1999; 399(6738):814-7.
29. Gong JG, Costanzo A, Yang HQ et al. The tyrosine kinase c-Abl regulates p73 in apoptotic response to cisplatin-induced DNA damage [see comments]. Nature 1999; 399(6738):806-9.
30. Agami R, Blandino G, Oren M et al. Interaction of c-Abl and p73alpha and their collaboration to induce apoptosis [see comments]. Nature 1999; 399(6738):809-13.
31. Shishido T, Akagi T, Chalmers A et al. Crk family adaptor proteins trans-activate c-Abl kinase. Genes Cells 2001; 6(5):431-40.
32. Comer AR, Ahern-Djamali SM, Juang JL et al. Phosphorylation of Enabled by the Drosophila Abelson tyrosine kinase regulates the in vivo function and protein-protein interactions of Enabled. Mol Cell Biol 1998; 18(1):152-60.
33. Gertler FB, Comer AR, Juang JL et al. enabled, a dosage-sensitive suppressor of mutations in the Drosophila Abl tyrosine kinase, encodes an Abl substrate with SH3 domain-binding properties. Genes Dev 1995; 9(5):521-33.
34. Gertler FB, Hill KK, Clark MJ et al. Dosage-sensitive modifiers of Drosophila abl tyrosine kinase function: Prospero, a regulator of axonal outgrowth, and disabled, a novel tyrosine kinase substrate [published erratum appears in Genes Dev 1996 Sep 1;10(17):2234]. Genes Dev 1993; 7(3):441-53.
35. Yu HH, Zisch AH, Dodelet VC et al. Multiple signaling interactions of Abl and Arg kinases with the EphB2 receptor. Oncogene 2001; 20(30):3995-4006.
36. Yano H, Cong F, Birge RB et al. Association of the Abl tyrosine kinase with the Trk nerve growth factor receptor. J Neurosci Res 2000; 59(3):356-64.
37. Zipfel PA, Grove M, Blackburn K et al. The c-Abl tyrosine kinase is regulated downstream of the B cell antigen receptor and interacts with CD19. J Immunol 2000; 165(12):6872-9.
38. Mayer BJ, Jackson PK, Van Etten RA et al. Point mutations in the abl SH2 domain coordinately impair phosphotyrosine binding in vitro and transforming activity in vivo. Mol Cell Biol 1992; 12:609-618.
39. Pellicena P, Stowell KR, Miller WT. Enhanced phosphorylation of Src family kinase substrates containing SH2 domain binding sites. J Biol Chem 1998; 273(25):15325-8.
40. Mayer BJ, Hirai H, Sakai R. Evidence that the SH2 domains promote processive phosphorylation by protein-tyrosine kinases. Curr Biol 1995; 5:296-305.
41. Duyster J, Baskaran R, Wang JY. Src homology 2 domain as a specificity determinant in the c-Abl-mediated tyrosine phosphorylation of the RNA polymerase II carboxyl-terminal repeated domain. Proc Natl Acad Sci USA 1995; 92(5):1555-9.

42. Zhou S, Carraway IIIrd KL, Eck MJ et al. Catalytic specificity of protein-tyrosine kinases is critical for selective signalling. Nature 1995; 373(6514):536-9.
43. Zhou S, Cantley LC. Recognition and specificity in protein tyrosine kinase-mediated signalling. Trends Biochem Sci 1995; 20(11):470-5.
44. Schindler T, Bornmann W, Pellicena P et al. Structural mechanism for STI-571 inhibition of abelson tyrosine kinase. Science 2000; 289(5486):1938-42.
45. Nolen B, Taylor S, Ghosh G. Regulation of protein kinases; controlling activity through activation segment conformation. Mol Cell 2004; 15(5):661-75.
46. Dorey K, Engen JR, Kretzschmar J et al. Phosphorylation and structure-based functional studies reveal a positive and a negative role for the activation loop of the c-Abl tyrosine kinase. Oncogene 2001; 20(56):8075-84.
47. Brasher BB, Van Etten RA. c-Abl has high intrinsic tyrosine kinase activity that is stimulated by mutation of the src homology 3 domain and by autophosphorylation at two distinct regulatory tyrosines. J Biol Chem 2000; 275(45):35631-7.
48. Plattner R, Kadlec L, DeMali KA et al. c-Abl is activated by growth factors and src family kinases and has a role in the cellular response to PDGF. Genes Dev 1999; 13(18):2400-11.
49. Plattner R, Irvin BJ, Guo S et al. A new link between the c-Abl tyrosine kinase and phosphoinositide signalling through PLC-gamma1. Nat Cell Biol 2003; 5(4):309-19.
50. Tanis KQ, Veach D, Duewel HS et al. Two distinct phosphorylation pathways have additive effects on abl family kinase activation. Mol Cell Biol 2003; 23(11):3884-96.
51. Reynolds Jr FH, Oroszlan S, Stephenson JR. Abelson murine leukemia virus P120: Identification and characterization of tyrosine phosphorylation sites. J Virol 1982; 44(3):1097-101.
52. Steen H, Fernandez M, Ghaffari S et al. Phosphotyrosine mapping in Bcr/Abl oncoprotein using phosphotyrosine-specific immonium ion scanning. Mol Cell Proteomics 2003; 2(3):138-45.
53. Salomon AR, Ficarro SB, Brill LM et al. Profiling of tyrosine phosphorylation pathways in human cells using mass spectrometry. Proc Natl Acad Sci USA 2003; 100(2):443-8.
54. Cong F, Spencer S, Cote JF et al. Cytoskeletal protein PSTPIP1 directs the PEST-type protein tyrosine phosphatase to the c-Abl kinase to mediate Abl dephosphorylation. Mol Cell 2000; 6(6):1413-23.
55. Echarri A, Pendergast AM. Activated c-Abl is degraded by the ubiquitin-dependent proteasome pathway. Curr Biol 2001; 11(22):1759-65.
56. Schlessinger J. Cell signaling by receptor tyrosine kinases. Cell 2000; 103(2):211-25.
57. McWhirter JR, Galasso DL, Wang JY. A coiled-coil oligomerization domain of Bcr is essential for the transforming function of Bcr-Abl oncoproteins. Mol Cell Biol 1993; 13(12):7587-95.
58. Zhao X, Ghaffari S, Lodish H et al. Structure of the Bcr-Abl oncoprotein oligomerization domain. Nat Struct Biol 2002; 7:117-20.
59. Golub TR, Goga A, Barker GF et al. Oligomerization of the ABL tyrosine kinase by the Ets protein TEL in human leukemia. Mol Cell Biol 1996; 16(8):4107-16.
60. Beissert T, Puccetti E, Bianchini A et al. Targeting of the N-terminal coiled coil oligomerization interface of BCR interferes with the transformation potential of BCR-ABL and increases sensitivity to STI571. Blood 2003; 102(8):2985-93.
61. Smith KM, Van Etten RA. Activation of c-Abl kinase activity and transformation by a chemical inducer of dimerization. J Biol Chem 2001; 276(26):24372-9.
62. Zhang X, Subrahmanyam R, Wong R et al. The NH(2)-terminal coiled-coil domain and tyrosine 177 play important roles in induction of a myeloproliferative disease in mice by Bcr-Abl. Mol Cell Biol 2001; 21(3):840-853.
63. Druker BJ. Imatinib as a paradigm of targeted therapies. Adv Cancer Res 2004; 91:1-30.
64. Azam M, Latek RR, Daley GQ. Mechanisms of autoinhibition and STI-571/Imatinib resistance revealed by mutagenesis of BCR-ABL. Cell 2003; 112(6):831-43.
65. Brasher BB, Roumiantsev S, Van Etten RA. Mutational analysis of the regulatory function of the c-Abl Src homology 3 domain. Oncogene 2001; 20(53):7744-52.
66. Roumiantsev S, Shah NP, Gorre ME et al. Clinical resistance to the kinase inhibitor STI-571 in chronic myeloid leukemia by mutation of Tyr-253 in the Abl kinase domain P-loop. Proc Natl Acad Sci USA 2002; 99(16):10700-5.
67. Wang JY. Controlling Abl: Auto-inhibition and co-inhibition? Nat Cell Biol 2004; 6(1):3-7.
68. Xu W, Doshi A, Lei M et al. Crystal structures of c-Src reveal features of its autoinhibitory mechanism. Mol Cell 1999; 3(5):629-38.
69. DeLano WL. The PyMOL molecular graphics system. http://pymol.sourceforge.net 2002.
70. Cong F, Spencer S, Cote J et al. Cytoskeletal protein PSTPIP1 directs the PEST-type protein tyrosine phosphatase to the c-Abl kinase to mediate Abl. Dephosphorylation 2000.

CHAPTER 2

Role of Abl Family Kinases in Growth Factor-Mediated Signaling

Ann Marie Pendergast*

Abstract

Constitutive activation of the Abl nonreceptor tyrosine kinases can occur as a result of structural alterations of the *Abl1* and *Abl2* genes, which leads to the development of leukemia. Tight control of the activities of the endogenous Abl kinases is critical for normal development and maintenance of normal homeostasis. Recent work has revealed that the Abl kinases are activated by diverse cell surface receptors, and function to couple the activated receptors to signaling pathways such as those important for the regulation of cytoskeletal dynamics.

The c-Abl (Abl1) tyrosine kinase was identified over two decades ago as a proto-oncoprotein deregulated in human, mouse and feline leukemias.[1] Only recently have the normal functions of c-Abl and the related Arg (Abl2) tyrosine kinases begun to be elucidated. Abl kinases have been shown to regulate cytoskeletal reorganization, cell migration, and responses to DNA damage and oxidative stress.

Whereas Abl kinases are activated constitutively as a result of structural alterations in leukemias, we showed that normal Abl kinases are transiently activated by growth factor stimulation. Both c-Abl and Arg are activated by stimulation of cells with platelet-derived growth factor (PDGF) or epidermal growth factor (EGF).[2,3] The mechanism of Abl kinase activation by growth factor receptors requires activation of two types of receptor-associated enzymes: Src kinases and phospholipase C-γ1 (PLC-γ1).[2-4] Src kinases phosphorylate Abl family proteins in the activation loop of the kinase domain, and at a second site in the inter-linker region between the kinase and SH2 domains.[2-4] Although Src kinases are necessary, they are not sufficient for activation of Abl kinases by PDGF (Fig. 1). We showed that PLC-γ1 is also required for PDGF-mediated activation of both c-Abl and Arg.[3,5] Our data revealed that the PLC-γ1 substrate, phosphatidyl inositol bisphosphate (PIP_2) inhibits the activity of Abl kinases in vitro and in cells,[2] and that decreasing cellular levels of PIP_2 by PLC-γ1-mediated hydrolysis produces dramatic activation of the endogenous Abl family kinases.[3,5] Our data is consistent with the notion that PIP_2 or a molecule bound to PIP_2 functions to stabilize the inactive conformation of the Abl kinases. Depletion of cellular levels of PIP_2 by PLC-γ1 releases inhibitory interactions and leads to Abl kinase activation, which is further augmented by direct phosphorylation by Src family kinases (Fig. 1).

More recently we showed that Abl family kinases are recruited to the PDGF receptor (PDGFR) within minutes after stimulation with the PDGF ligand, and that the Abl kinases are

*Coresponding Author: Ann Marie Pendergast—Department of Pharmacology and Cancer Biology, Duke University Medical Center, Durham, North Carolina, U.S.A., Email: pende014@mc.duke.edu

Abl Family Kinases in Development and Disease, edited by Anthony Koleske.
©2006 Landes Bioscience and Springer Science+Business Media.

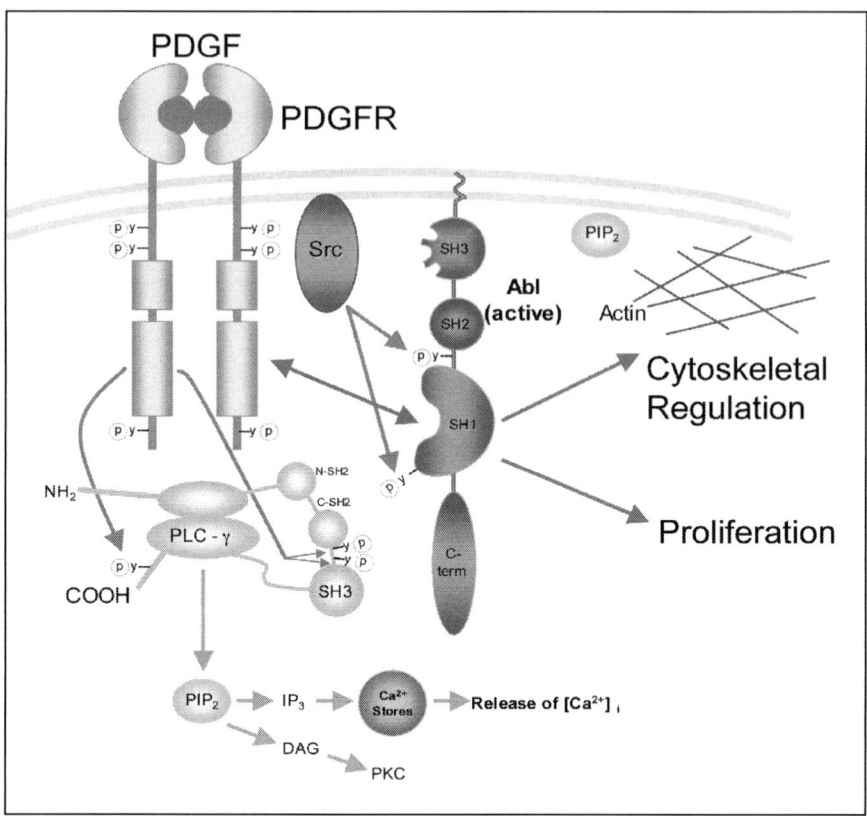

Figure 1. Activation of Abl Kinases by PDGF Signaling. Stimulation of the PDGF receptor (PDGFR) results in tyrosine phosphorylation of the cytoplasmic tail of the receptor and recruitment of SH2-containing signaling proteins such as Src and PLC-γ1 to specific tyrosine phosphorylated sites. Recruitment and activation of both Src and PLC-γ1 is required for maximal activation of Abl family kinases by PDGF. Activation of Abl kinases requires hydrolysis of PIP2 by PLC-γ1 and tyrosine phosphorylation of Abl kinases by Src. Activation of Abl kinases downstream of the PDGFR plays a role in the transmission of signals leading to cytoskeletal reorganization and cell proliferation.

phosphorylated by the activated PDGFR.[3] The Abl kinases bind to the tyrosine phosphorylated PDGFR via the Abl SH2 domain. The Abl kinases in turn phosphorylate the PDGFR (Fig. 1). The functional consequences of the reciprocal phosphorylation of Abl kinases and the PDGFR remain to be determined. Nevertheless, we have shown that activation of c-Abl by PDGF is physiologically significant, as mouse embryo fibroblasts (MEFs) derived from c-Abl knockout mice are deficient in their ability to undergo cytoskeletal reorganization in response to PDGF compared to wild-type control MEFs.[2] This deficit is rescued by reexpression of wild-type c-Abl in the null MEFs.[2] Subsequent work by Scita et al demonstrated that MEFs derived from Abl/Arg-double knockout mice are markedly deficient in their ability to form ruffles in re-sponse to PDGF and that this defect correlates with decreased Rac activation in cells lacking c-Abl and Arg.[6] Moreover, we showed that c-Abl functions downstream of PLC-γ1 in the regulation of chemotaxis towards PDGF in endothelial cells and fibroblasts.[3,5] Unexpectedly, we observed that while reexpression of c-Abl in Abl/Arg-null MEFs rescues the ability of PLC-γ1 to increase PDGF-mediated chemotaxis, reexpression of Arg in these cells fails to rescue the chemotaxis defect.[3] Thus, although both c-Abl and Arg kinases are activated downstream of

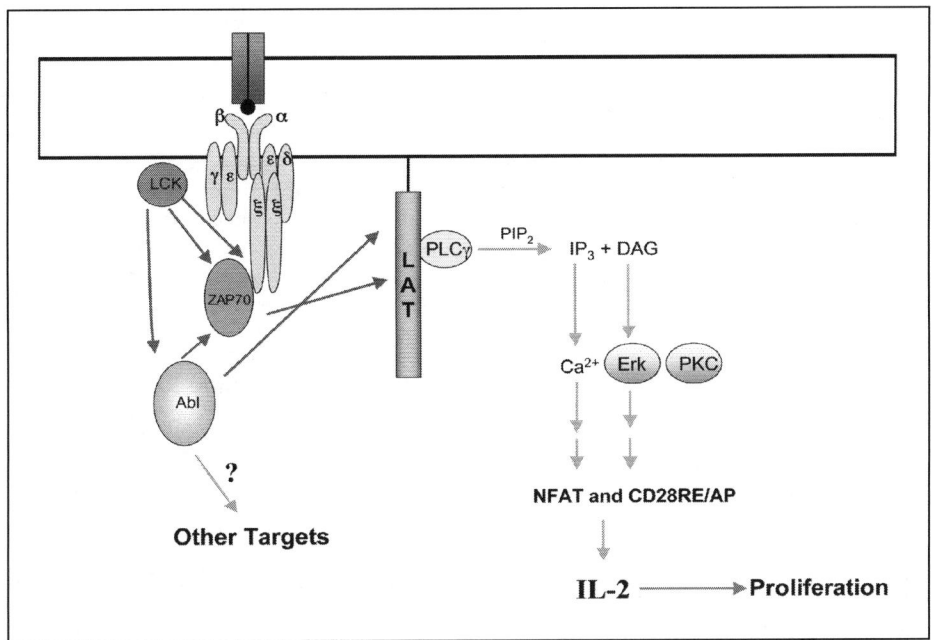

Figure 2. Role of Abl Kinases downstream of T cell receptor (TCR) stimulation. Activation of the TCR by antigen results in activation of the Lck tyrosine kinase. This leads to phosphorylation of the ITAM of the TCRζ chain and the recruitment and phosphorylation of the Zap70 tyrosine kinase. Abl kinases are activated by TCR engagement, and this activation requires Lck. Activation of the Abl kinases leads to phosphorylation of Zap70 on tyrosine 319, and also results in enhanced tyrosine phosphorylation of the LAT adaptor, which leads to recruitment and enhanced tyrosine phosphorylation of PLC-γ1. Hydrolysis of PIP2 by PLC-γ1 induces activation of protein kinase C, Erk, and calcium-dependent pathways leading to activation of NFAT and CD28RE/AP elements of the IL-2 promoter. Enhanced IL-2 production is critical for T cell proliferation. Activation of Abl kinases is likely to affect other pathways through phosphorylation of other targets in antigen-stimulated T cells.

the PDGFR, they have nonredundant function during PLC-γ1-dependent chemotaxis towards PDGF.

In addition to their role in the regulation of cytoskeletal responses downstream of the activated PDGFR, we and others found that Abl kinases also function to regulate PDGF-dependent mitogenesis.[2,7,8] Cells that are null for c-Abl exhibit delayed S-phase entry after PDGF stimulation compared to the same cells reconstituted with physiological levels of c-Abl.[2,8] Similarly, c-Abl/Arg-deficient fibroblasts exhibit decreased PDGF-induced proliferation and function downstream of Src in this pathway.[7]

Recruitment and activation of Abl family kinases by receptor kinases has now been extended to the MuSK and Eph receptor tyrosine kinases.[9,10] Abl kinases are activated following stimulation of MuSK by agrin, and in turn the activated Abl kinases phosphorylate and activate the MuSK receptor kinase.[9] We showed that Abl kinases are required for MuSK-dependent clustering of acetylcholine receptors.[9] This finding revealed a novel post-synaptic role for Abl family kinases as direct post-synaptic signaling components downstream of MuSK at the neuromuscular junction.

In addition to activation by receptor tyrosine kinases, Abl kinases are also activated by cell surface receptors lacking kinase activity such as the transmembrane-type Semaphorin 6D (Sema 6D). Activation of Sema 6D by the Plexin A1 receptor leads to recruitment of the Abl protein to the cytoplasmic tail of Sema 6D, and subsequent Abl kinase activation.[11] Abl kinases appear to have a role downstream of Sema 6D for the induction of cardiac cell migration into the trabeculae.[11] Abl kinases have also been linked to adhesion receptors such as the Netrin receptor Frazzled and the Roundabout (Robo) receptors.[12,13] However, activation of Abl kinases by engagement of these receptors has not yet been demonstrated.

Homozygous loss of both c-Abl and Arg in mice results in embryonic lethality.[14] The mice most likely die of hemorrhage and present elevated numbers of apoptotic cells in all tissues of the body. The embryos from mice lacking both c-Abl and Arg exhibit neurulation defects, which may be linked to alterations in the actin cytoskeleton. The pleitropic phenotypes of Abl/Arg-double knockout mice may be due to interrupted signaling linking diverse cell surface receptors to pathways critical for the regulation of cytoskeletal dynamics, cell survival and proliferation.

Among the most striking phenotypes of c-Abl knockout mice are a variety of immune phenotypes, including splenic and thymic atrophy, lymphopenia, and increased susceptibility to infection.[15,16] Only recently have the molecular mechanisms responsible for these phenotypes been revealed. We showed that engagement of the T cell receptor (TCR) results in activation of the endogenous Abl kinases and that this activation is dependent in part on the activation of the Src family kinase Lck[17] (Fig. 2). The activation of Abl kinases by the TCR is reminiscent of the activation of Abl kinases by growth factor receptor tyrosine kinases such as the PDGFR, which also requires Src family kinase activity.[2,3] Following activation of the Abl kinases in response to TCR engagement, the Abl kinases phosphorylate the Zap70 tyrosine kinase and the transmembrane adaptor linker for activation of T cells (LAT)[17] (Fig. 2). Loss of Abl kinase activity by treatment with a pharmacological inhibitor or genetic ablation of Abl and Arg in primary T cells results in markedly decreased tyrosine phosphorylation of LAT and Zap70 on tyrosine 319.[17] Significantly, we showed that Abl kinases are required for cell proliferation and IL-2 production in response to TCR stimulation of primary T cells.[17] Similarly, B cells derived from c-Abl null or mutant mice exhibit reduced proliferation following stimulation of the B cell receptor (BCR).[18,19] Thus, Abl kinases play a role in the activation of signaling pathways downstream of immunoreceptors in T and B cells, and function to modulate cell proliferation, and possibly other cellular responses in these cells (Fig. 2). These findings are consistent with the immunological phenotypes of Abl-deficient mice, and more recently, the observed immunosupression in patients treated with a pharmacological inhibitor of the Abl kinases.[20]

In summary, it is now increasingly evident that Abl kinases play critical roles downstream of cell surface receptors in the regulation of cell proliferation, survival, adhesion and motility. Future work is likely to reveal additional roles for the Abl kinases during development and in pathological conditions.

Acknowledgements

I thank present and past members of the Pendergast lab for their important contributions towards advancing our understanding of the function of the Abl family of tyrosine kinases. In particular, I thank Dr. Patricia Zipfel for her outstanding intellectual and experimental contributions. I also thank Dr. Zipfel for figure preparation. The work in my laboratory was supported by NIH grants CA70940, AI056266, and NS050392.

References

1. Pendergast AM. The Abl family kinases: Mechanisms of regulation and signaling. Adv Cancer Res 2002; 85:51-100.
2. Plattner R, Kadlec L, DeMali KA et al. C-Abl is activated by growth factors and Src family kinases and has a role in the cellular response to PDGF. Genes and Dev 1999; 13:2400-2411.
3. Plattner R, Koleske AJ, Kazlauskas A et al. Bidirectional signaling links the Abelson kinases to the platelet-derived growth factor receptor. Mol Cell Biol 2004; 24:2573-2583.
4. Tanis KQ, Veach D, Duewel HS et al. Two distinct phosphorylation pathways have additive effects on Abl family kinase activation. Mol Cell Biol 2003; 23:3884-3896.

5. Plattner R, Irvin BJ, Guo S et al. A new link between the c-Abl tyrosine kinase and phosphoinositide signaling through PLC-γ1. Nat Cell Biol 2003; 5:309-319.
6. Sini P, Cannas A, Koleske AJ et al. Abl-dependent tyrosine phosphorylation of Sos-1 mediates growth-factor-induced Rac activation. Nat Cell Biol 2004; 6:268-274.
7. Furstoss O, Dorey K, Simon V et al. C-Abl is an effector of Src for growth factor-induced c-myc expression and DNA synthesis. EMBO J 2002; 21:514-524.
8. Plattner R, Pendergast AM. Activation and signaling of the Abl tyrosine kinase. Cell Cycle 2003; 2:A8-A9.
9. Finn AJ, Feng G, Pendergast AM. Postsynaptic requirement for Abl kinases in assembly of the neuromuscular junction. Nat Neurosci 2003; 6:717-723.
10. Yu HH, Zisch AH, Dodelet VC et al. Multiple signaling interactions of Abl and Arg kinases with the EphB2 receptor. Oncogene 2001; 20:3995-4006.
11. Toyofuku T, Zhang H, Kumanogoh A et al. Guidance of myocardial patterning in cardiac development by Sema 6D reverse signaling. Nat Cell Biol 2004; 6:1204-1211.
12. Forsthoefel DJ, Liebl EC, Kolodziej PA et al. The Abelson tyrosine kinase, the trio GEF and Enabled interact with the Netrin receptor Frazzled in Drosophila. Development 2005; 132:1983-1994.
13. Rhee J, Mahfooz NS, Arregui C et al. Activation of the repulsive receptor Roundabout inhibits N-cadherin-mediated cell adhesion. Nat Cell Biol 2002; 4:798-805.
14. Koleske AJ, Gifford AM, Scott ML et al. Essential roles for the Abl and Arg tyrosine kinases in neurulation. Neuron 1998; 21:1259-1272.
15. Tybulewicz VL, Crawford CE, Jackson PK et al. Neonatal lethality and lymphopenia in mice with a homozygous disruption of the c-abl proto-oncogene. Cell 1991; 65:1153-1163.
16. Schwartzberg PL, Stall AM, Hardin JD et al. Mice homozygous for the abl[m1] mutation show poor viability and depletion of selected B and T cell populations. Cell 1991; 65:1165-1175.
17. Zipfel PA, Zhang W, Quiroz M et al. Requirement for Abl kinases in T cell receptor signaling. Current Biol 2004; 14:1222-1231.
18. Zipfel PA, Grove M, Blackburn K et al. The c-Abl tyrosine kinase is regulated downstream of the B cell antigen receptor and interacts with CD19. J Immunol 2000; 165:6872-6879.
19. Hardin JD, Boast S, Schwartzberg PL et al. Abnormal peripheral lymphocyte function in c-abl mutant mice. Cell Immunol 1996; 172:100-107.
20. Wange RL. TCR signaling: Another Abl-bodied kinase joins the cascade. Current Biol 2004; 14:R562-R564.

CHAPTER 3

Regulation of Cell Adhesion Responses by Abl Family Kinases

Keith Quincy Tanis* and Martin Alexander Schwartz

Abstract

Integrins are cell surface receptors that mediate the interactions of cells with each other and the extracellular matrix. In this chapter, we review experiments indicating that the Abl family of nonreceptor tyrosine kinases, Abl and Arg in vertebrates, are important mediators of cellular responses to integrin engagement. During the early stages of cell spreading, integrins trigger the activation of Abl family kinases and their association with multiple focal adhesion proteins. These events lead to phosphorylation of several cytoskeletal regulatory proteins and changes in cell morphology and motility. Integrins may also utilize Abl family kinases to regulate nuclear processes such as gene expression, cell cycle progression and cell survival. Defects in the proper modulation of cell adhesive responses by Abl family kinases are thought to contribute to the progression of chronic myelogenous leukemia and could potentially underlie other human diseases and behavioral disorders.

Introduction

Cells live in a meshwork of proteins and polysaccharides known as the extracellular matrix (ECM). The ECM directs many aspects of localized cell behavior such as proliferation, differentiation, migration and polarity that are required for tissue organization and function. The interactions of cells with the ECM as well as with neighboring cells are mediated through several classes of cell surface adhesion molecules including integrins, cadherins, immunoglobulins, proteoglycans, and selectins.

Integrins are the principle receptors for binding most ECM proteins such as fibronectin, collagen, vitronectin, and laminin.[1-4] Integrins are heterodimers consisting of one α- and one β-subunit. Eighteen different α-subunits and eight different β-subunits have been identified and are known to combine into 24 αβ heterodimers.[4,5] Each integrin recognizes a specific set of ECM proteins and cell-surface ligands.[4] Ligand binding to integrins results in changes in integrin conformation and clustering, and recruitment of numerous signaling and cytoskeletal proteins.[3,5-7] The resulting aggregates of ECM proteins, integrins, cytoskeletal proteins and signaling molecules can form several different types of adhesions including focal complexes, focal adhesions, podosomes or fibrillar adhesions.[8] These integrin-mediated adhesions comprise hubs where adhesion, signaling, cytoskeletal reorganization, and mechanical stresses all interact to drive diverse functions including cell shape, polarity, motility, survival, proliferation, and differentiation.[1,3,7,9]

*Corresponding Author: Keith Quincy Tanis—Department of Psychiatry, Laboratory of Molecular Psychiatry, Yale University, 34 Park Street, New Haven, Connecticut 06508 U.S.A. Email: keith.tanis@yale.edu

Abl Family Kinases in Development and Disease, edited by Anthony Koleske.
©2006 Landes Bioscience and Springer Science+Business Media.

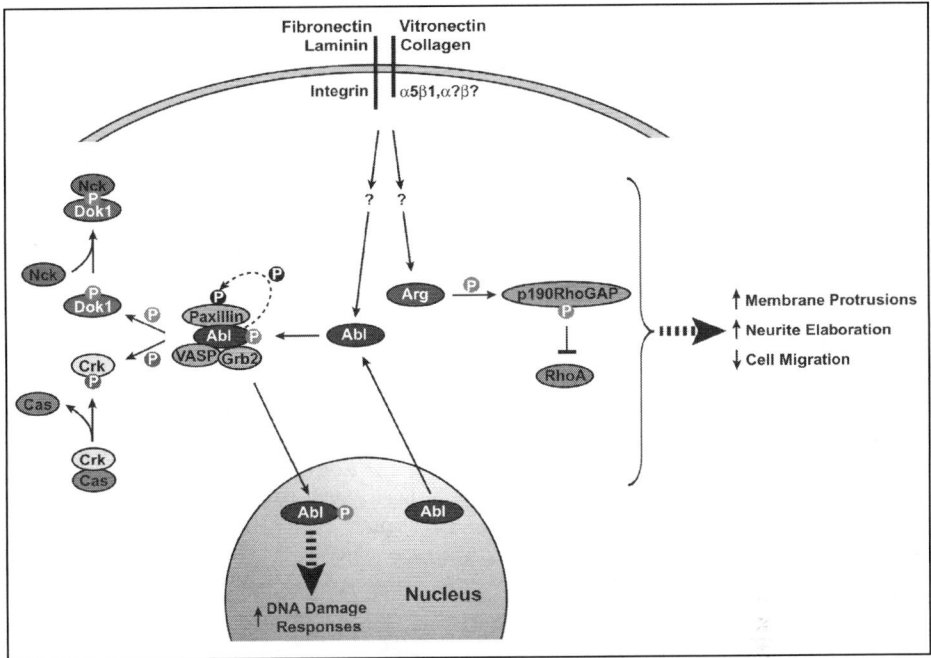

Figure 1. Regulation of Abl family kinase signaling by integrin receptors. Following integrin engagement, Abl is recruited from the nucleus to focal adhesions, is phosphorylated and activated, and associates with multiple proteins including paxillin, VASP and Grb2. Following activation, Abl phosphorylates Dok1 to promote its association with Nck, and phosphorylates Crk to disrupt Crk-Cas complexes. Abl may also phosphorylate paxillin upon integrin engagement. Integrins activate Arg to phosphorylate the Rho inhibitor, p190RhoGAP. Through these and yet unidentified interactions, Abl family kinases produce adhesion-dependent changes in cytoskeletal dynamics to induce membrane protrusions and neurite elaboration while inhibiting cell migration. As cells stably adhere, Abl returns to the nucleus, activated and primed to mediate DNA damage responses.

Research conducted over the last decade has revealed that the Abl family nonreceptor tyrosine kinases, Abl and Arg in vertebrates, are important mediators of integrin signaling (Fig. 1). Integrins regulate the subcellular localization, binding interactions and kinase activity of Abl family kinases to evoke both cytoskeletal and nuclear responses to cell adhesion. In this chapter, we review the currently known connections between Abl family kinases and cell adhesion responses.

The Bcr-Abl Fusion Protein Causes Defects in Integrin Function

The first link between Abl family kinases and cell adhesion came from studies of the Bcr-Abl fusion protein. Bcr-Abl is expressed as a result of a chromosomal translocation between chromosomes 9 and 22, and is the causative agent of nearly all cases of chronic myelogenous leukemia (CML).[10-12] Expression of Bcr-Abl causes multiple abnormalities in cell adhesion and integrin signaling pathways.[13,14] Hematopoietic progenitors from CML patients exhibit decreased adhesion to stromal layers and fibronectin but increased adhesion to laminin and collagen type IV.[15-18] These changes have been hypothesized to mediate premature release of Bcr-Abl cells into the circulation. However, several groups have observed increased cell adhesion to fibronectin when Bcr-Abl is introduced into hematopoietic cell lines.[13,19-21] These differences likely result from differences in cell type, length of adhesion, cytokines and culture conditions.

Bcr-Abl transformed cells also exhibit enhanced and persistent motility on fibronectin-coated surfaces.[22] While adhesion to fibronectin inhibits cell cycle progression in normal hematopoietic cells, this regulation is lost in Bcr-Abl transformed cells.[23,24] Conversely, adhesion to ECM proteins is normally required for DNA proliferation in fibroblasts, but this requirement is relieved by expression of Bcr-Abl without diminishing the requirement for growth factors or serum.[25] Overall, the disruption of integrin function by Bcr-Abl is suspected to contribute to the abnormal trafficking and expansion of CML progenitors.

Several properties of Bcr-Abl contribute to the disruption of normal integrin signaling. Fusion of Bcr to Abl hyperactivates Abl kinase activity and results in the phosphorylation of several focal adhesion proteins including paxillin, FAK, vinculin, talin, tensin, CRKL, p130Cas and Cbl.[26-30] Constitutive phosphorylation of at least some of these proteins by Bcr-Abl can interfere with proper signaling in response to integrin engagement. For example, Bcr-Abl induces constitutive association of CRKL with paxillin and p130Cas, and disrupts the association of p130Cas and tensin.[30,31] In addition to its elevated tyrosine kinase activity, direct cytoskeletal interactions mediated by Abl's C-terminal filamentous (F)-actin binding domain contribute to the effects of Bcr-Abl on cell adhesion.[18,32] Bcr-Abl's effects on adhesion also require the Bcr coiled-coil domain,[32] most likely because this domain mediates oligomerization of Bcr-Abl to enhance Abl kinase activity and the association of Bcr-Abl with actin filaments.[33,34] Finally, deletion of the Abl proline-rich region in Bcr-Abl reduces the adhesion defects caused by Bcr-Abl, suggesting that interactions between Bcr-Abl and SH3-domain containing proteins are involved.[18]

Abl Family Kinases Mediate Cellular Responses to Integrin Engagement

Many of the interactions between Bcr-Abl and focal adhesion components likely result from changes in localization, binding partners, substrate affinities, and kinase activity caused by the fusion of Bcr and Abl. However, these observations raised the question whether normal Abl family kinases are involved in physiological integrin signaling. Indeed, genetic, pharmacological, and biochemical analyses have revealed that Abl family kinases are important mediators of both the cytoskeletal and nuclear responses to integrin engagement (Fig. 1).

Cytoskeletal Responses

Abl stimulates the formation of actin microspikes and Arg promotes lamellipodial protrusions and retractions in fibroblasts spreading on fibronectin.[35,36] Abl family kinases also modulate neuronal morphology in response to integrin engagement. When cultured on the integrin ligand laminin-1, cortical neurons from both *abl*[-/-] mice and *arg*[-/-] mice exhibit reduced neurite outgrowth and branching relative to wild type neurons.[37,86] In contrast, *arg*[-/-] and wild type neurons elaborate neurites similarly when cultured on poly-ornithine, a substrate that does not engage integrin receptors.[86] Further, laminin-1 mediated neurite outgrowth and branching in wild type neurons is prevented either by inhibiting β1 integrins with echistatin or Abl family kinases with STI571.[86] Together, these experiments indicate that Abl family kinases are required for integrin-mediated neurite outgrowth and branching, and that they contribute to the regulation of protrusive structures in fibroblasts adhering to ECM proteins.

Abl family kinases also regulate fibroblast motility on integrin ligands. Deletion of Abl or inhibition of Abl and Arg kinase activity with STI571 results in enhanced cell migration on fibronectin, while over-expression of Abl inhibits fibroblast migration on fibronectin.[38,39] This inhibitory effect was linked to phosphorylation of a negative regulatory tyrosine on Crk.[38] Phosphorylation of this site prevents Crk from binding to p130Cas and thereby decreases Rac activation and cell migration.[38,40]

Nuclear Responses

Although Arg is confined to the cytoplasm, Abl localizes to both the cytoplasm and the nucleus.[41-44] In the nucleus, Abl is involved in cellular responses to DNA damage, cell cycle progression and apoptosis.[45,46] Abl is activated by DNA damage in adherent cells but not in suspended cells.[47] Decreased Abl activity in suspended cells correlates with stabilization of the p53 homolog p73 and decreased apoptosis after DNA damage.[47] These data reveal a surprising synergy between integrin pathways and the nuclear pathways that regulate the response to DNA damage.

Abl Family Kinases Are Recruited to Sites of Adhesive Contact

Abl is found in the nucleus and diffusely in the cytoplasm in stably adhered or suspended fibroblasts.[41,42] However, during fibroblast attachment to the ECM proteins fibronectin, vitronectin, or collagen, Abl exits the nucleus and colocalizes with integrins at newly formed sites of adhesion (Fig. 1).[42] Levels of Abl in the nucleus are lowest 20 minutes after adhesion to ECM proteins and then slowly return to the levels observed in stably attached cells.[42] These effects are specific to integrin-mediated adhesion as fibroblast adhesion to poly-L-lysine does not induce Abl export from the nucleus or localization to sites of adhesion.[42]

The mechanisms that trigger integrin-mediated export of Abl from the nucleus and the localization of Abl to focal adhesions have not been determined. As described below, Abl associates with several focal adhesion proteins upon integrin engagement and these interactions may recruit Abl to focal adhesions. It remains to be determined whether Arg, like Abl, is recruited to sites of focal contact.

Integrin Engagement Activates Abl Family Kinases

Abl kinase activity, measured in immunoprecipitates, declines about 3-fold upon detachment of stably adherent fibroblasts.[42] Upon reattachment to fibronectin or to an antibody against integrin α_5, Abl kinase activity transiently increases 4- to 5-fold and then returns to the plateau level observed in stably adherent cells.[42] Fibroblast adhesion to fibronectin also stimulates Arg kinase activity with similar magnitude and kinetics as Abl (K.Q.T. and Anthony J. Koleske, unpublished). In contrast, no change in Abl[42] or Arg (K.Q.T. and Anthony J. Koleske, unpublished) kinase activity is observed when fibroblasts adhere to poly-L-lysine. Further, integrin-mediated phosphorylation of the Arg substrate p190RhoGap is prevented by STI571 or by genetic deletion of Arg in fibroblasts or neurons.[48] Together, these experiments indicate that the kinase activities of both Abl and Arg are activated upon integrin engagement by ECM proteins (Fig. 1).

As mentioned earlier, Abl returns to the nucleus at later times after adhesion.[42] While enhanced Abl kinase activity is observed in the cytoplasmic pool within 5 minutes of cell adhesion to fibronectin, Abl kinase activity in the nucleus does not increase until after 20 minutes of cell adhesion, which parallels the transport of Abl back to the nucleus.[42] Together, these data suggest that upon integrin engagement, nuclear Abl is exported to the cytoplasm where it is activated and then returned to the nucleus. This mechanism may allow Abl to relay signals from integrin receptors directly to the nucleus to mediate integrin regulation of gene expression, cell cycle progression, differentiation or survival.

It is poorly understood how integrins activate Abl and Arg. A recent study suggested that Abl and Arg activation by adhesion may be mediated in part by the release from inhibitory interactions with F-actin.[49] F-actin was found to inhibit Abl kinase activity in vitro, and deletion of Abl's F-actin binding domain enhanced the in vivo kinase activity of Abl and reduced its adhesion-dependence.[49] On the other hand, adhesion to fibronectin has been shown to stimulate Abl phosphorylation.[50] Although the sites of phosphorylation have not been determined, Abl and Arg contain multiple phosphorylation sites that activate their kinase activities.[51,52] Adhesion-dependent phosphorylation of Abl requires Abl kinase activity, implying either that it results from autophosphorylation or that Abl kinase activity is needed to recruit another kinase to phosphorylate Abl.[50] However, it remains to be determined whether phosphorylation mediates adhesion-dependent activation of Abl.

Tyrosine kinases play a key role in relaying signals from integrin receptors. The focal adhesion kinase (FAK) is activated during integrin-mediated adhesion.[53] Src kinases, which phosphorylate and activate Abl and Arg following stimulation of growth factor receptors, are recruited to activated FAK where they promote phosphorylation of numerous proteins including the Abl substrates, paxillin and Cas.[52,54-56] PLC-γ1 also contributes to the activation of Abl and Arg by growth factors, and is recruited and activated by FAK upon integrin engagement.[56-58] Further studies are needed to determine whether FAK, Src or PLC-γ1 participate in the activation of Abl and Arg by integrin receptors.

The specific integrin receptors that regulate Abl and Arg also remain to be carefully determined. An antibody that crosslinks α_5 integrins activates Abl kinase activity, suggesting that the $\alpha_5\beta_1$ fibronectin receptor can activate Abl.[42] Other data implicate Abl and Arg in neuronal responses to laminin-1 which binds multiple integrins including $\alpha_1\beta_1$ and $\alpha_6\beta_1$ in neurons (Eva M.Y. Moresco and Anthony J. Koleske, unpublished).[37,59] Abl localization to adhesion sites is observed in fibroblasts plated on collagen and vitronectin, which implicates integrins $\alpha_2\beta_1$ and $\alpha_v\beta_3$.[42] Thus, Abl family kinases are most likely regulated by multiple integrin receptors, but detailed analysis of kinase regulation by these receptors remains to be carried out. Determining the spectrum of integrin receptors and their ligands that regulate Abl and Arg will give important insights into the physiological processes that utilize the cell adhesion responses mediated by Abl family kinases.

Abl Family Kinases Interact with Multiple Proteins upon Integrin Engagement

Upon fibroblast adhesion to ECM proteins, Abl family kinases are known to associate with multiple proteins and to phosphorylate several substrates (Fig 1). These interactions are described below and likely contribute to the ability of Abl family kinases to regulate the cytoskeletal and nuclear responses to integrin engagement.

Paxillin

Paxillin is an adapter protein that localizes to focal adhesions and modulates cell motility and gene expression.[60,61] Plating suspended fibroblasts on fibronectin stimulates transient co-immunoprecipitation of paxillin and Abl.[50] This association is lost at later times of adhesion.

Immunoprecipitated Abl can phosphorylate paxillin in vitro, and its ability to do so is enhanced following cell adhesion to fibronectin.[50] It is not known which paxillin tyrosines are phosphorylated by Abl, whether Abl or Arg phosphorylate paxillin in vivo, or what physiological consequences these phosphorylation events may have. However, as tyrosine phosphorylation of paxillin recruits multiple proteins to focal adhesions and regulates cell motility, Abl phosphorylation of paxillin has the potential to regulate these events.[60,61]

VASP

The vasodilator-stimulated phosphoprotein (VASP) regulates actin polymerization and localizes to focal adhesions and to the leading edge of membrane protrusions.[62] Abl co-immunoprecipitates with VASP in adherent cells, and this association is enhanced during the early stages of cell spreading when Abl kinase activity is elevated.[63] Although the consequences of this interaction remain to be determined, VASP may localize Abl to adhesive structures and/or to membrane protrusions, and may mediate the membrane protrusive effects of Abl and Arg.

Grb2

Another protein that co-immunoprecipitates with Abl specifically during cell spreading is Grb2, an adaptor protein that couples cell-surface receptors to MAP kinase signaling.[64] Integrin-mediated activation of MAP kinase signaling regulates cell-cycle progression and migration.[65-67] Over-expression of kinase-inactive Abl decreases the activation of Erk2 upon integrin engagement, suggesting that Abl may contribute to integrin activation of MAP kinase signaling.[64] Abl induces phosphorylation of the Grb2-associated guanine nucleotide-exchange factor Sos-1 during growth factor stimulation, resulting in Rac activation and membrane ruffles.[68] This suggests that Abl may also contribute to the activation of Sos-1 and Rac during integrin engagement, although this hypothesis remains to be tested.

Dok1

Dok1 (down stream of tyrosine kinases) is hyper-phosphorylated on tyrosine in Bcr-Abl transformed cells and was recently found to be an in vivo substrate of Abl in wild type fibroblasts.[36,69-71] Abl mediated phosphorylation of Dok1 is enhanced by adhesion to fibronectin, and is required for Abl-induced filopodial extensions during cell spreading.[36] Abl/Dok1-mediated filopodial extensions require Nck, an adaptor protein that under some circumstances can induce localized actin polymerization.[36,72,73] Phosphorylation of Dok1 by Abl stimulates the association of Dok1 and Nck.[36] Together, these experiments suggest that upon activation by integrin receptors, Abl phosphorylates Dok1 to promote its association with Nck and the formation of filopodial extensions.

p190RhoGap

The 190-KDa GTPase activating protein for Rho (p190RhoGAP) was recently identified as an Arg substrate.[48] RhoA promotes the formation of focal adhesions and actin stress fibers.[74] Phosphorylation of p190RhoGAP upon fibroblast adhesion to ECM proteins increases its GAP activity towards RhoA.[75] This results in transient RhoA inactivation to promote efficient cell spreading.[75] Integrin-induced phosphorylation of p190RhoGAP and inactivation of Rho was previously found to be mediated by Src and FAK.[76,77] However, Arg is also required for integrin-mediated p190RhoGAP phosphorylation in both fibroblasts and neurons.[48] The relationship between Arg and FAK/Src in this pathway remains to be investigated. Arg mediated inactivation of Rho contributes to the formation of membrane protrusions, and may explain the increased actin stress fibers observed in *arg*[−/−] fibroblasts.[35,44,48]

Crk

Although it has not been shown explicitly, Abl and Arg are likely to also phosphorylate the Crk adapter protein following activation by integrins. Both Crk and Abl associate with paxillin following integrin engagement, potentially positioning Crk to be phosphorylated by Abl.[50,78] As mentioned previously, Abl phosphorylates Crk at an inhibitory tyrosine that disrupts the Crk-p130Cas complex and decreases cell migration.[38,40]

Lasp-1

Cell adhesion or exposure to growth factors induces the translocation of the actin binding protein Lasp-1 from the cell periphery to focal adhesions.[79] Abl was recently reported to phosphorylate Lasp-1 upon activation by DNA damage or oxidative stress, and phosphorylation of Lasp-1 by Abl was found to prevent its translocation to focal adhesions.[79] This effect is specific to stress pathways, since activation of Abl by cell adhesion or growth factors did not result in detectable phosphorylation of Lasp-1 or prevent Lasp-1 relocalization.[79] In this study, loss of Lasp-1 from focal adhesions was linked to increased apoptosis in response to stresses, suggesting that Abl promotes cell death partly by inhibiting an integrin-dependent survival pathway mediated by Lasp-1.

Together, these studies indicate that integrins modulate the interactions of Abl family kinases with multiple proteins that regulate focal adhesion signaling and cytoskeletal dynamics. It should be noted that these identified interactions represent only a few pieces to a complex puzzle. Additional proteins and signaling events likely contribute to the adhesive responses mediated by Abl family kinases.

Concluding Remarks and Future Challenges

The findings discussed in this chapter clearly indicate that Abl family kinases are functionally relevant mediators of cellular responses to integrin engagement. During the early stages of cell spreading, integrins trigger the association of Abl family kinases with multiple focal adhesion proteins and elevate the kinase activities of Abl and Arg to promote the phosphorylation of several cytoskeletal regulatory proteins. By modulating cytoskeletal dynamics in response to integrin engagement, Abl and Arg produce adhesion dependent changes in cell morphology and motility. Integrins may also utilize Abl family kinases to regulate nuclear processes such as gene expression, cell cycle progression and cell survival. However, much remains to be learned about the biochemical mechanisms whereby Abl family kinases link integrin receptors to downstream cellular responses and about the physiological consequences of these signaling events.

A comprehensive understanding of the cell adhesion responses mediated by Abl family kinases and their biological significance requires the completion of several future challenges. The first challenge is to determine how Abl family kinase signaling is regulated by adhesive cues. This includes determining the specific integrin receptors that regulate Abl and Arg and the biochemical mechanisms by which they transiently induce Abl and Arg relocalization, kinase activation, and signaling protein interactions. It should also be determined how integrins coordinate with other stimuli to regulate Abl and Arg. It will be particularly interesting to determine how cell adhesion regulates the activation of Abl by DNA damage and whether growth factor and integrin receptors cooperate to produce additive enhancements of Abl and Arg signaling cascades. The second challenge is identify the complete list of Abl and Arg signaling interactions regulated by adhesion and the biochemical mechanisms whereby these interactions produce downstream cellular responses. The third challenge is to understand the physiological consequences of these cellular responses and how they impact the function of tissues.

The last and most important challenge is to determine how defects in the cell adhesion responses of Abl family kinases contribute to human disease. As discussed above, improper regulation of integrin signaling cascades by Bcr-Abl likely contributes to the abnormal expansion and trafficking of CML progenitors. The inability to properly respond to adhesive cues may also explain some of the defects observed in mice lacking Abl or Arg.[80-82] For example, mice lacking Arg exhibit multiple behavior abnormalities such as reduced mating and aggression that may result from the reduced dendrite arborization observed in these mice (Eva M.Y. Moresco and Anthony J. Koleske, unpublished data).[82] Reduced dendrite arbors are also observed in several human disorders such as mood disorders, and mental retardation syndromes.[83-85] More work is required to determine whether defects in Abl family kinase signaling contribute to these or other human disorders.

Acknowledgments

We would like to thank Tony Koleske, Bill Bradley, Scott Boyle, and Stefanie Lapetina for helpful comments on this manuscript. This work was supported by National Institutes of Health grants RO1 47214 (M.A.S.) and MH67388 (K.Q.T.).

References

1. Hynes RO. Integrins: versatility, modulation, and signaling in cell adhesion. Cell 1992; 69(1):11-25.
2. Schwartz MA, Schaller MD, Ginsberg MH. Integrins: emerging paradigms of signal transduction. Annu Rev Cell Dev Biol 1995; 11:549-599.
3. Giancotti FG, Ruoslahti E. Integrin signaling. Science 1999; 285(5430):1028-1032.

4. Plow EF, Haas TA, Zhang L et al. Ligand binding to integrins. J Biol Chem 2000; 275(29):21785-21788.
5. van der Flier A, Sonnenberg A. Function and interactions of integrins. Cell Tissue Res 2001; 305(3):285-298.
6. Hynes RO. Integrins: bidirectional, allosteric signaling machines. Cell 2002; 110(6):673-687.
7. Miranti CK, Brugge JS. Sensing the environment: a historical perspective on integrin signal transduction. Nat Cell Biol 2002; 4(4):E83-90.
8. Zamir E, Geiger B. Molecular complexity and dynamics of cell-matrix adhesions. J Cell Sci 2001; 114(Pt 20):3583-3590.
9. Schwartz MA, Ginsberg MH. Networks and crosstalk: integrin signalling spreads. Nat Cell Biol 2002; 4(4):E65-68.
10. Nowell PC, Hungerford DA. Chromosome studies on normal and leukemic human leukocytes. J Natl Cancer Inst 1960; 25:85-109.
11. Rowley JD. Letter: A new consistent chromosomal abnormality in chronic myelogenous leukaemia identified by quinacrine fluorescence and Giemsa staining. Nature 1973; 243(5405):290-293.
12. Bartram CR, de Klein A, Hagemeijer A et al. Translocation of c-abl oncogene correlates with the presence of a Philadelphia chromosome in chronic myelocytic leukaemia. Nature 1983; 306(5940):277-280.
13. Salesse S, Verfaillie CM. Mechanisms underlying abnormal trafficking and expansion of malignant progenitors in CML: BCR/ABL-induced defects in integrin function in CML. Oncogene 2002; 21(56):8605-8611.
14. Wertheim JA, Miller JP, Xu L et al. The biology of chronic myelogenous leukemia:mouse models and cell adhesion. Oncogene 2002; 21(56):8612-8628.
15. Gordon MY, Dowding CR, Riley GP et al. Altered adhesive interactions with marrow stroma of haematopoietic progenitor cells in chronic myeloid leukaemia. Nature 1987; 328(6128):342-344.
16. Verfaillie CM, McCarthy JB, McGlave PB. Mechanisms underlying abnormal trafficking of malignant progenitors in chronic myelogenous leukemia. Decreased adhesion to stroma and fibronectin but increased adhesion to the basement membrane components laminin and collagen type IV. J Clin Invest 1992; 90(4):1232-1241.
17. Bhatia R, Munthe HA, Verfaillie CM. Role of abnormal integrin-cytoskeletal interactions in impaired beta1 integrin function in chronic myelogenous leukemia hematopoietic progenitors. Exp Hematol 1999; 27(9):1384-1396.
18. Ramaraj P, Singh H, Niu N et al. Effect of mutational inactivation of tyrosine kinase activity on BCR/ABL-induced abnormalities in cell growth and adhesion in human hematopoietic progenitors. Cancer Res 2004; 64(15):5322-5331.
19. Bazzoni G, Carlesso N, Griffin JD et al. Bcr/Abl expression stimulates integrin function in hematopoietic cell lines. J Clin Invest 1996; 98(2):521-528.
20. Kramer A, Horner S, Willer A et al. Adhesion to fibronectin stimulates proliferation of wild-type and bcr/abl-transfected murine hematopoietic cells. Proc Natl Acad Sci USA 1999; 96(5):2087-2092.
21. Wertheim JA, Forsythe K, Druker BJ et al. BCR-ABL-induced adhesion defects are tyrosine kinase-independent. Blood 2002; 99(11):4122-4130.
22. Salgia R, Li JL, Ewaniuk DS et al. BCR/ABL induces multiple abnormalities of cytoskeletal function. J Clin Invest 1997; 100(1):46-57.
23. Hurley RW, McCarthy JB, Verfaillie CM. Direct adhesion to bone marrow stroma via fibronectin receptors inhibits hematopoietic progenitor proliferation. J Clin Invest 1995; 96(1):511-519.
24. Lundell BI, McCarthy JB, Kovach NL et al. Activation-dependent alpha5beta1 integrin-mediated adhesion to fibronectin decreases proliferation of chronic myelogenous leukemia progenitors and K562 cells. Blood 1996; 87(6):2450-2458.
25. Renshaw MW, McWhirter JR, Wang JY. The human leukemia oncogene bcr-abl abrogates the anchorage requirement but not the growth factor requirement for proliferation. Mol Cell Biol 1995; 15(3):1286-1293.
26. Oda T, Heaney C, Hagopian JR et al. Crkl is the major tyrosine-phosphorylated protein in neutrophils from patients with chronic myelogenous leukemia. J Biol Chem 1994; 269(37):22925-22928.
27. Gotoh A, Miyazawa K, Ohyashiki K et al. Tyrosine phosphorylation and activation of focal adhesion kinase (p125FAK) by BCR-ABL oncoprotein. Exp Hematol 1995; 23(11):1153-1159.
28. Salgia R, Brunkhorst B, Pisick E et al. Increased tyrosine phosphorylation of focal adhesion proteins in myeloid cell lines expressing p210BCR/ABL. Oncogene 1995; 11(6):1149-1155.
29. Salgia R, Sattler M, Pisick E et al. p210BCR/ABL induces formation of complexes containing focal adhesion proteins and the protooncogene product p120c-Cbl. Exp Hematol 1996; 24(2):310-313.

30. Salgia R, Pisick E, Sattler M et al. p130CAS forms a signaling complex with the adapter protein CRKL in hematopoietic cells transformed by the BCR/ABL oncogene. J Biol Chem 1996; 271(41):25198-25203.
31. Salgia R, Uemura N, Okuda K et al. CRKL links p210BCR/ABL with paxillin in chronic myelogenous leukemia cells. J Biol Chem 1995; 270(49):29145-29150.
32. Wertheim JA, Perera SA, Hammer DA et al. Localization of BCR-ABL to F-actin regulates cell adhesion but does not attenuate CML development. Blood 2003; 102(6):2220-2228.
33. McWhirter JR, Wang JY. Activation of tyrosinase kinase and microfilament-binding functions of c-abl by bcr sequences in bcr/abl fusion proteins. Mol Cell Biol 1991; 11(3):1553-1565.
34. McWhirter JR, Galasso DL, Wang JY. A coiled-coil oligomerization domain of Bcr is essential for the transforming function of Bcr-Abl oncoproteins. Mol Cell Biol 1993; 13(12):7587-7595.
35. Miller AL, Wang Y, Mooseker MS et al. The Abl-related gene (Arg) requires its F-actin-microtubule cross-linking activity to regulate lamellipodial dynamics during fibroblast adhesion. J Cell Biol 2004; 165(3):407-419.
36. Woodring PJ, Meisenhelder J, Johnson SA et al. c-Abl phosphorylates Dok1 to promote filopodia during cell spreading. J Cell Biol 2004; 165(4):493-503.
37. Woodring PJ, Litwack ED, O'Leary DD et al. Modulation of the F-actin cytoskeleton by c-Abl tyrosine kinase in cell spreading and neurite extension. J Cell Biol 2002; 156(5):879-892.
38. Kain KH, Klemke RL. Inhibition of cell migration by Abl family tyrosine kinases through uncoupling of Crk-CAS complexes. J Biol Chem 2001; 276(19):16185-16192.
39. Frasca F, Vigneri P, Vella V et al. Tyrosine kinase inhibitor STI571 enhances thyroid cancer cell motile response to Hepatocyte Growth Factor. Oncogene 2001; 20(29):3845-3856.
40. Feller SM, Knudsen B, Hanafusa H. c-Abl kinase regulates the protein binding activity of c-Crk. EMBO J 1994; 13(10):2341-2351.
41. Van Etten RA, Jackson P, Baltimore D. The mouse type IV c-abl gene product is a nuclear protein, and activation of transforming ability is associated with cytoplasmic localization. Cell 1989; 58(4):669-678.
42. Lewis JM, Baskaran R, Taagepera S et al. Integrin regulation of c-Abl tyrosine kinase activity and cytoplasmic- nuclear transport. Proc Natl Acad Sci USA 1996; 93(26):15174-15179.
43. Wang B, Mysliwiec T, Krainc D et al. Identification of ArgBP1, an Arg protein tyrosine kinase binding protein that is the human homologue of a CNS-specific Xenopus gene. Oncogene 1996; 12(9):1921-1929.
44. Wang Y, Miller AL, Mooseker MS et al. The Abl-related gene (Arg) nonreceptor tyrosine kinase uses two F-actin- binding domains to bundle F-actin. Proc Natl Acad Sci USA 2001; 98(26):14865-14870.
45. Van Etten RA. Cycling, stressed-out and nervous: cellular functions of c-Abl. Trends Cell Biol 1999; 9(5):179-186.
46. Pendergast AM. The Abl family kinases: mechanisms of regulation and signaling. Adv Cancer Res 2002; 85:51-100.
47. Truong T, Sun G, Doorly M et al. Modulation of DNA damage-induced apoptosis by cell adhesion is independently mediated by p53 and c-Abl. Proc Natl Acad Sci USA 2003; 100(18):10281-10286.
48. Hernandez SE, Settleman J, Koleske AJ. Adhesion-dependent regulation of p190RhoGAP in the developing brain by the Abl-related gene tyrosine kinase. Curr Biol 2004; 14(8):691-696.
49. Woodring PJ, Hunter T, Wang JY. Inhibition of c-Abl tyrosine kinase activity by filamentous actin. J Biol Chem 2001; 276(29):27104-27110.
50. Lewis JM, Schwartz MA. Integrins regulate the association and phosphorylation of paxillin by c-Abl. J Biol Chem 1998; 273(23):14225-14230.
51. Brasher BB, Van Etten RA. c-Abl has high intrinsic tyrosine kinase activity that is stimulated by mutation of the Src homology 3 domain and by autophosphorylation at two distinct regulatory tyrosines. J Biol Chem 2000; 275(45):35631-35637.
52. Tanis KQ, Veach D, Duewel HS et al. Two distinct phosphorylation pathways have additive effects on Abl family kinase activation. Mol Cell Biol 2003; 23(11):3884-3896.
53. Parsons JT. Focal adhesion kinase: the first ten years. J Cell Sci 2003; 116(Pt 8):1409-1416.
54. Cary LA, Guan JL. Focal adhesion kinase in integrin-mediated signaling. Front Biosci 1999; 4:D102-113.
55. Plattner R, Kadlec L, DeMali KA et al. c-Abl is activated by growth factors and Src family kinases and has a role in the cellular response to PDGF. Genes Dev 1999; 13(18):2400-2411.
56. Plattner R, Koleske AJ, Kazlauskas A et al. Bidirectional signaling links the Abelson kinases to the platelet-derived growth factor receptor. Mol Cell Biol 2004; 24(6):2573-2583.

57. Zhang X, Chattopadhyay A, Ji QS et al. Focal adhesion kinase promotes phospholipase C-gamma1 activity. Proc Natl Acad Sci USA 1999; 96(16):9021-9026.
58. Plattner R, Irvin BJ, Guo S et al. A new link between the c-Abl tyrosine kinase and phosphoinositide signalling through PLC-gamma1. Nat Cell Biol 2003; 5(4):309-319.
59. Clegg DO, Wingerd KL, Hikita ST et al. Integrins in the development, function and dysfunction of the nervous system. Front Biosci 2003; 8:d723-750.
60. Schaller MD. Paxillin: a focal adhesion-associated adaptor protein. Oncogene 2001; 20(44):6459-6472.
61. Brown MC, Turner CE. Paxillin: adapting to change. Physiol Rev 2004; 84(4):1315-1339.
62. Kwiatkowski AV, Gertler FB, Loureiro JJ. Function and regulation of Ena/VASP proteins. Trends Cell Biol 2003; 13(7):386-392.
63. Howe AK, Hogan BP, Juliano RL. Regulation of vasodilator-stimulated phosphoprotein phosphorylation and interaction with Abl by protein kinase A and cell adhesion. J Biol Chem 2002; 277(41):38121-38126.
64. Renshaw MW, Lewis JM, Schwartz MA. The c-Abl tyrosine kinase contributes to the transient activation of MAP kinase in cells plated on fibronectin. Oncogene 2000; 19(28):3216-3219.
65. Assoian RK, Schwartz MA. Coordinate signaling by integrins and receptor tyrosine kinases in the regulation of G1 phase cell-cycle progression. Curr Opin Genet Dev 2001; 11(1):48-53.
66. Juliano RL, Reddig P, Alahari S et al. Integrin regulation of cell signalling and motility. Biochem Soc Trans 2004; 32(Pt3):443-446.
67. Slack-Davis JK, Parsons JT. Emerging views of integrin signaling: implications for prostate cancer. J Cell Biochem 2004; 91(1):41-46.
68. Sini P, Cannas A, Koleske AJ et al. Abl-dependent tyrosine phosphorylation of Sos-1 mediates growth-factor-induced Rac activation. Nat Cell Biol 2004; 6(3):268-274.
69. Wisniewski D, Strife A, Wojciechowicz D et al. A 62-kilodalton tyrosine phosphoprotein constitutively present in primary chronic phase chronic myelogenous leukemia enriched lineage negative blast populations. Leukemia 1994; 8(4):688-693.
70. Carpino N, Wisniewski D, Strife A et al. p62(dok): a constitutively tyrosine-phosphorylated, GAP-associated protein in chronic myelogenous leukemia progenitor cells. Cell 1997; 88(2):197-204.
71. Yamanashi Y, Baltimore D. Identification of the Abl- and rasGAP-associated 62 kDa protein as a docking protein, Dok. Cell 1997; 88(2):205-211.
72. Campellone KG, Rankin S, Pawson T et al. Clustering of Nck by a 12-residue Tir phosphopeptide is sufficient to trigger localized actin assembly. J Cell Biol 2004; 164(3):407-416.
73. Rivera GM, Briceno CA, Takeshima F et al. Inducible clustering of membrane-targeted SH3 domains of the adaptor protein Nck triggers localized actin polymerization. Curr Biol 2004; 14(1):11-22.
74. Ridley AJ, Hall A. The small GTP-binding protein rho regulates the assembly of focal adhesions and actin stress fibers in response to growth factors. Cell 1992; 70(3):389-399.
75. Arthur WT, Noren NK, Burridge K. Regulation of Rho family GTPases by cell-cell and cell-matrix adhesion. Biol Res 2002; 35(2):239-246.
76. Arthur WT, Petch LA, Burridge K. Integrin engagement suppresses RhoA activity via a c-Src-dependent mechanism. Curr Biol 2000; 10(12):719-722.
77. Ren XD, Kiosses WB, Sieg DJ et al. Focal adhesion kinase suppresses Rho activity to promote focal adhesion turnover. J Cell Sci 2000; 113(Pt 20):3673-3678.
78. Schaller MD, Parsons JT. pp125FAK-dependent tyrosine phosphorylation of paxillin creates a high-affinity binding site for Crk. Mol Cell Biol 1995; 15(5):2635-2645.
79. Lin YH, Park ZY, Lin D et al. Regulation of cell migration and survival by focal adhesion targeting of Lasp-1. J Cell Biol 2004; 165(3):421-432.
80. Tybulewicz VL, Crawford CE, Jackson PK et al. Neonatal lethality and lymphopenia in mice with a homozygous disruption of the c-abl proto-oncogene. Cell 1991; 65(7):1153-1163.
81. Li B, Boast S, de los Santos K et al. Mice deficient in Abl are osteoporotic and have defects in osteoblast maturation. Nat Genet 2000; 24(3):304-308.
82. Koleske AJ, Gifford AM, Scott ML et al. Essential roles for the Abl and Arg tyrosine kinases in neurulation. Neuron 1998; 21(6):1259-1272.
83. D'Sa C, Duman RS. Antidepressants and neuroplasticity. Bipolar Disord 2002; 4(3):183-194.
84. Quiroz JA, Singh J, Gould TD et al. Emerging experimental therapeutics for bipolar disorder: clues from the molecular pathophysiology. Mol Psychiatry 2004; 9(8):756-76.
85. Kaufmann WE, Moser HW. Dendritic anomalies in disorders associated with mental retardation. Cereb Cortex 2000; 10(10):981-991.
86. Moresco EYM, Donaldson S, Williamson A et al. Intergrin-mediated dendrite branch maintenance requires Abelson (Abl) family kinase. J Neurosci 20004; 25(26):6105-6118.

CHAPTER 4

Abl and Cell Death

Jean Y.J. Wang,* Yosuke Minami and Jiangyu Zhu

Abstract

The Abl tyrosine kinase contains nuclear-import and -export signals and undergoes nucleo-cytoplasmic shuttling in proliferating cells. The nuclear Abl is activated by DNA damage or tumor necrosis factor to promote cell death through transcription-dependent and -independent mechanisms. The oncogenic BCR-ABL tyrosine kinase is defective in nuclear import and functions as an inhibitor of apoptosis in the cytoplasm. If allowed to function in the nucleus, BCR-ABL also induces cell death. Abl interacts with several different types of death effectors. However, the precise mechanism by which Abl tyrosine kinase regulates cell death remains to be determined.

Introduction

The mammalian Abl tyrosine kinase is ubiquitously expressed: Abl mRNA and protein are found in all tissues and cell types examined, from embryonic stem cells to mature spermatids. This nonreceptor tyrosine kinase has been shown to function as a transducer of a variety of cell extrinsic and intrinsic signals including those from growth factors, cell adhesion, inflammatory cytokines, oxidative stress, and DNA damage.[1-3] Activated Abl tyrosine kinase regulates cytoskeletal function, cell cycle progression, myogenic differentiation and cell death. This diverse array of biological activities is dictated by several modular functional domains, which determine the subcellular localization and the interaction partners of Abl (Fig. 1).

The N-terminal region of Abl resembles the Src-family of tyrosine kinases in that it contains the Src-homology (SH) 3, 2, and tyrosine kinase domains. These N-terminal domains assemble through intra-molecular interactions into an auto-inhibited conformation with low catalytic activity.[4-6] The C-terminal region of Abl, not found in Src-family members, contains three nuclear localization signals (NLS), one nuclear export signal (NES), binding sites for G-actin and F-actin as well as binding site for double stranded A/T-rich DNA (Fig. 1). Several cellular proteins, including the retinoblastoma protein (RB) and F-actin that interact with the kinase domain or the C-terminus respectively, can further inhibit Abl kinase activity by enforcing the auto-inhibited conformation.[5,7] Yet other cellular proteins binding through the SH3 domain or the proline-rich linker (PRL) in the C-terminal region of Abl can activate its kinase activity.[2,3,7,8] Therefore, formation and disruption of Abl interactions with inhibitory or stimulatory cellular proteins underlie the regulation of this tyrosine kinase.[5,6]

The accumulated observations on Abl have suggested it to have a wide range of biological activities (discussed in the companion chapters of this book). We have previously proposed that Abl protein may be partitioned into different signaling complexes, each of which subjects Abl kinase to regulation by a specific signal.[1,2] The partitioning of Abl may also affect its

*Corresponding Author: Jean Y.J. Wang—Division of Hematology-Oncology, Department of Medicine, Moores Cancer Center, School of Medicine, University of California San Diego, La Jolla, California 92093, U.S.A. Email: jywang@ucsd.edu

Abl Family Kinases in Development and Disease, edited by Anthony Koleske.
©2006 Landes Bioscience and Springer Science+Business Media.

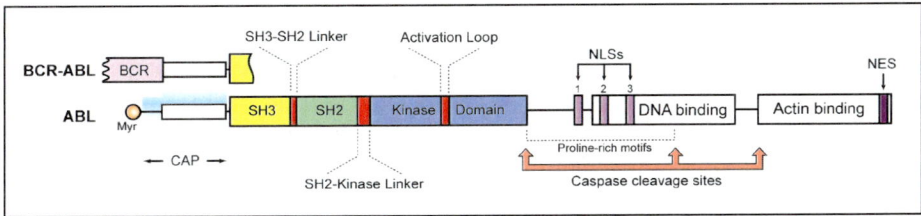

Figure 1. Schematics of Abl functional domains. The human ABL (1a) and murine Abl (type IV) is myristoylated at the N-terminus. Intramolecular interactions among the myristoyl-group, the CAP, the SH3 domain, the SH3-SH2 linker, the SH2 domain, the SH2-kinase linker and the kinase domain have been demonstrated by crystallography to establish an auto-inhibited conformation that inactivates Abl kinase. This auto-inhibited conformation is disrupted in BCR-ABL fusion protein through the loss of myristoylation and part of the CAP. In addition, an N-terminal coiled-coil in BCR causes the formation of BCR-ABL oligomers, further enhancing auto-phosphorylation and the constitutive activation of kinase activity. The C-terminal region of ABL contains subcellular location cues, including three nuclear localization signals (NLSs) and a nuclear export signal (NES). In the cytoplasm, ABL associates with actin. In the nucleus, ABL associates with DNA. A number of cellular proteins directly interact with ABL through its SH3, SH2, kinase domains, the proline-rich motifs or the actin-binding domain to regulate its kinase activity or to serve as its substrates. Three caspase cleavage sites are found in the C-terminal region of ABL. These cleavage events would disrupt the nucleo-cytoplasmic shuttling of ABL. The cell death regulatory function of ABL is critically dependent on its subcellular localization.

biological output, which would be dictated by locally available substrates and/or downstream effectors. This "partitioning" model for Abl function can accommodate the many biological functions that have been linked to Abl and explain the collection of pleiotropic and incompletely penetrant defects associated with *Abl*-knockout mice. While the idea of partitioning Abl maybe comforting, it is not satisfying. The selective agglomeration of the Abl functional domains in one protein, in theory, should have a higher purpose than to simply assign this tyrosine kinase to a disparate array of functions. The cumulative information has suggested two prominent biological themes for Abl function: the first is on actin dynamics and the second is in the regulation of cell death. In this chapter, we will focus the discussion on Abl's role in regulating programmed cell death.

Abl in Genotoxin-Induced Cell Death

DNA Damage Activates Abl Kinase

The mammalian Abl tyrosine kinase is activated by a variety of physical or chemical agents that damage the cellular DNA. Ionizing radiation, cytarabine (AraC), methyl methanesulfonate (MMS), mitomycin C (MMC), cisplatin, camptothecin, etoposide and doxorubicin have all been found to stimulate Abl kinase.[1,9-12] However, UV irradiation does not appear to increase the kinase activity of Abl.[9] Activation of Abl kinase by genotoxic agents has been observed in fibroblasts, fibrosarcoma cells, thymocytes, lymphoblasts, and cell lines derived from colon, breast and liver cancers.

The increase in Abl activity is detected by immune complex kinase assay, in which Abl is immunoprecipitated from cell lysates and then used to phosphorylate recombinant substrates purified from bacteria as GST-fusion proteins, including GST-CTD (C-terminal repeated domain of mammalian RNA polymerase II),[13-15] GST-Crk,[16-18] and GST-CrkCTD.[7,19] With this kinase assay, a genotoxin-induced increase in activity is detectable with Abl from nuclear extracts but not from cytoplasmic or total cellular extracts.[9,20] A 3-fold increase in Abl kinase measured by such an assay could be due a general 3-fold increase in the catalytic activity of every Abl molecule or a 100-fold increase in 3% of the Abl. This uncertainty renders "negative" results, i.e., lack of increase in Abl kinase activity, difficult to interpret. It is formally possible that genotoxic agents may activate a small fraction of the cytoplasmic pool of Abl to a level that is not detectable by the immune complex kinase assay. Despite this uncertainty, activation of nuclear pool of Abl tyrosine kinase by genotoxic agents has been reproducibly observed.

The mechanism by which genotoxic agents activate the nuclear Abl kinase has not been fully elucidated. It has been shown that nuclear Abl tyrosine kinase cannot be activated by ionizing radiation in ATM-deficient human and mouse cells.[21,22] ATM is a member of PI3/PI4 kinase family. Its mutation underlies *ataxiatelangiectasia* (A-T)—the genetic disease predisposing to cancer.[23] The ATM kinase is activated by ionizing radiation (IR) and plays an essential role in orchestrating cellular responses to IR.[24,25] The fact that IR cannot activate the nuclear Abl tyrosine kinase in ATM-deficient cells places Abl downstream of ATM in the DNA damage signal transduction pathway. IR-induced activation of Abl is abolished by the mutation of a single serine residue (Ser465) in the Abl kinase domain.[22] Although the Abl-Ser465Ala protein has kinase activity that is activated by cell adhesion,[22] it is not activated by IR or cisplatin, suggesting this phosphorylation to be a general requirement for Abl activation by genotoxins. Ser465 of Abl is located in the C-lobe of the kinase domain and situated in a characteristic SQ motif that is preferentially phosphorylated by ATM and its related PIKK.[21,22] Phosphorylation of Abl by ATM has been observed in vitro. However, in vivo phosphorylation of Abl at Ser465 by ATM has yet to be demonstrated.

Activation of nuclear Abl tyrosine kinase by cisplatin requires not only ATM but also a functional mismatch repair system, as cisplatin does not activate Abl kinase in MLH1-deficient human colon cancer cells.[26] Mismatch repair proteins, including MSH2, MSH6, MLH1, and PMS2, are involved not only in DNA repair but also in DNA damage signal transduction.[27,28] The MSH2/MSH6 heterodimer can bind to platinum-adducts in DNA.[29,30] Together with MLH1 and PMS2, these MMR proteins may recruit proteins such as ATR and ATM to initiate DNA damage signal transduction. This model would explain the requirement for both MLH1 and ATM in activating nuclear Abl tyrosine kinase in response to cisplatin.

Abl Interaction with DNA Repair Proteins

In addition to ATM and MLH1, Abl interacts with several other proteins involved in DNA repair and damage signal transduction, including DNA-PK, BRCA1, RAD9, RAD51, RAD52, WRN, UV-DDB1, and TOP1 (Table 1).

DNA-PK is another member of the PIKK-family of protein kinases that are activated by DNA damage. DNA-PK plays a critical role in nonhomologous end joining repair of double stranded breaks. Recent results have implicated DNA-PK in apoptotic response to DNA damage.[31-33] DNA-PK is not required for Abl activation by IR.[34] It appears that Abl can modulate the kinase activity of DNA-PK through as yet undefined mechanism and with unknown biological consequences.

It has been proposed that Abl participates in DNA repair, because it can interact with several proteins and enzymes involved in double-stranded break repair (DNA-PK, BRCA1, RAD51, RAD52). If Abl were involved in DNA repair, it would not likely to play an essential role because *Abl*-knockout cells do not exhibit the characteristic phenotype of repair-deficient cells, i.e., hypersensitivity to ionizing radiation or other genotoxins that induce double stranded breaks. We have measured clonogenic survival of *Abl*-knockout cells as a function of radiation dose repeatedly without observing any indication of hypersensitivity. These observations suggest Abl to be dispensable for DNA repair and its interaction with repair proteins may serve to transduce "repair" signal rather than regulate the repair process.

Abl in DNA Damage-Induced Apoptosis

Activation of the Abl tyrosine kinase by DNA damage has been linked in mammalian cells to the induction of apoptosis. In human cancer cell lines, activation of Abl kinase can be correlated with the activation of p53 and its related p73 proteins, leading to the stimulation of apoptosis in response to ionizing radiation (IR) or chemotherapeutic agents (e.g., cisplatin).[26,35-37]

Table 1. ABL-interacting proteins relevant to the control of cell death

Name*	Gene ID*	Molecular Function	Function Relative to ABL	Ref.
ATM	472	Protein kinase belonging to the PI-3 kinase super-family involved in DNA damage signal transduction	Phosphorylation and activation of ABL kinase	21,22, 120,121
ATR	545	ATM-related protein kinase involved in DNA damage signal transduction and DNA replication	Interaction with BCR-ABL appears to impair ATR function	122
BRCA1	672	A nuclear scaffolding protein with E3-ligase activity and involved in DNA repair and DNA damage signal transduction	Interaction with ABL and phosphorylation of its C-terminus by ABL kinase	123
DDB1	1643	The large subunit of a heterodimeric DNA damage-binding protein involved in the repair of UV-adducts in DNA	Phosphorylation by ABL kinase to suppress UV-DDB activity	124
MDM2	4193	E3-Ubiquitin ligase required for the degradation of p53	Phosphorylation by ABL kinase to neutralize its degradation of TP53	65,66
NCSTN (APH2)	23385	A protein with DHHC zinc finger domain	Interaction with and co-localization of ABL to ER to induce cell death	91
POLR2A	5430	RNA polymerase II	Phosphorylation of its C-terminal domain (CTD) by ABL kinase to regulate transcription	14,15, 74,125
PRKDC (DNAPK)	5591	The catalytic subunit of DNA-dependent protein protein kinase (DNA-PK) belonging to the PI-3 kinase super-family, involved in non-homologous end joining repair of double stranded breaks.	Phosphorylation by ABL linked to inhibition of DNA-PK activity	34,126, 127
RAD51	5888	Catalyzes strand evasion during homologous recombination, similar to E. coli RecA and Saccharomyces cerevisiae Rad51.	Phosphorylation by ABL kinase correlates with its nuclear foci formation	128,129
RAD52	5893	A cofactor of Rad51 in the strand invasion reaction during homologous recombination	Phosphorylation by ABL kinase correlates with its nuclear foci formation	130,131
RAD9A	5883	A subunit of the 9-1-1 complex that resembles the homo-trimeric PCNA DNA clamp. It also possesses 3′-5′ exonuclease. The 9-1-1 complex is involved in the recognition of damaged DNA to stimulate repair and signaling pathways	Phosphorylation by ABL kinase at its BH3 domain enhances the binding of the BH3 domain to Bcl-xL	85

Continued on next Page

Table 1. Continued

Name*	Gene ID*	Molecular Function	Function Relative to ABL	Ref.
RB1	5925	The retinoblastoma tumor suppressor protein, a nuclear scaffolding protein involved in transcription regulation.	Binding to ABL kinase domain and inhibits ABL activity	45,50, 132-135
TOP1	7150	Topoisomerase I	Phosphorylation and activation by ABL kinase	136
TP53	7157	A transcription factor and a tumor suppressor	Phosphorylation by ABL kinase to increase its transcriptional activity	67,137
TP73	7161	A transcription factor belonging to the p53-family	Phosphorylation by ABL kinase to increase protein stability and transcription activity	26,35,36
WRN	7486	A member of the RecQ subfamily and the DEAH (Asp-Glu-Ala-His) subfamily of DNA and RNA helicases, loss of WRN causes premature aging.	Phosphorylation by ABL kinase to inhibit its exonulcease and helicase activities	138
NFKB IA	4792	Inhibitor of NF-kB that retains NF-kB in the cytoplasm	Phosphorylation by ABL kinase to induce its nuclear accumulation and inhibition of NF-κB	76
PRDX1	5052	A member of the peroxiredoxin family of antioxidant enzymes, which reduce hydrogen peroxide and alkyl hydroperoxides.	Oxidative stress-induced protein inhibitor of ABL kinase	139

*Nomenclatures (in alphabetical order) and IDs of the corresponding human genes are given in conformation to the Entrez Gene database of NCBI.

Fibroblasts derived from *Abl*-knockout mice showed reduced apoptotic response to several genotoxins, including cisplatin, camptothecin, cytosine arabinose and doxorubicin.[38,39] Mouse embryo fibroblasts do not undergo apoptosis in response to IR; hence, the contribution of Abl to IR-induced apoptosis could not be assessed in this experimental system. We have observed reduced apoptotic response to IR with *Abl*-knockout mouse thymocytes (Wood LD, Wang JYJ, unpublished observation). However, IR-induced apoptosis of neuroblasts in the developing central nervous system is not reduced in *Abl*-null mice although this response is dependent on ATM and p53.[40]

Taken together, the accumulated evidence supports the conclusion that activation of Abl by genotoxins can lead to the stimulation of apoptosis. However, Abl's pro-apoptotic function appears to be dependent on the type of genotoxins and the cellular context. In this regard, the Abl of *C. elegans* does not contribute to IR-induced apoptosis. In *C. elegans*, IR only induces apoptosis in germ cells, and this response is dependent on the worm p53.[41] Knockout of worm Abl sensitized *C. elegans* germ cells to IR,[42] suggesting Abl to protect these cells from IR-induced apoptosis. The Abl-mediated protection could be due either to its ability to stimulate DNA repair or its ability to inhibit p53-dependent apoptosis. At present, it is not known if worm Abl also interacts with DNA repair proteins such as RAD51 and RAD52. Because mammalian Abl must be activated in the nucleus to stimulate apoptosis (see below) and the worm Abl appears to be exclusively localized to the cytoplasm,[42] protection by Abl of apoptosis in IR-treated *C. elegans* germ cells may represent the cytoplasmic function of Abl.

Abl in TNF-Induced Cell Death

Activation of Abl by TNF

Tumor necrosis factor (TNF) is an inflammatory cytokine that coordinates the systemic responses to infection and injury. The type-I receptor for TNF (TNFRI) is ubiquitously expressed in all mammalian cell types. Upon stimulation by TNF, this receptor activates several signal transduction pathways to regulate a diverse array of cellular processes including programmed cell death. Activation of apoptosis by TNF is mediated through a caspase-dependent signal transduction pathway, in which TNFRI activates initiator caspases (caspase-8 and 10) to cleave downstream targets, either the effector caspases (caspase-3, 6, 7) or Bcl-family proteins to induce mitochondrial leakage of cytochrome C and AIF.[43,44] With cultured mammalian cells, stimulation with TNF alone is not sufficient to activate initiator caspases. TNF-induced apoptosis is generally induced by the combined treatment with TNF and cycloheximide (CHX).

Treatment of mouse fibroblasts or human U937 myeloid leukemia cells with TNF/CHX causes a measurable increase in the activity of nuclear Abl tyrosine kinase.[45,46] In U937 cells, TNF/CHX also activates JNK.[46] It was shown that Abl and JNK activation by TNF/CHX was independent of each other in these leukemia cells.[46] Interestingly, Abl activation by TNF/CHX does not occur when caspase is inhibited. Therefore, Abl activation is downstream of caspase activation in TNF/CHX-activated pathways. We have found that Abl tyrosine kinase activation by TNF/CHX requires the cleavage/degradation of the retinoblastoma protein, RB.[45] The RB protein is cleaved and degraded in apoptotic cells in a caspase-dependent manner.[47,48] Caspase can cleave RB at several sites. Mutation of a C-terminal caspase site in RB can prevent its cleavage/degradation; and preservation of this caspase-resistant RB-MI protein reduces the apoptotic response to TNF/CHX.[49] The caspase-site mutation has been introduced into the mouse *Rb-1* allele to create the *Rb-MI* mice.[49] The *Rb-MI* mice show increased resistance to endotoxin, which induces excessive inflammation to cause septic shock.[49] The intestinal epithelial cells of *Rb-MI* mice show reduced apoptosis during endotoxic shock.[49] Fibroblasts derived from *Rb-MI* mouse embryos also show reduced apoptotic response to TNF/CHX under conditions when TNFRI is stimulated.[49] While TNF/CHX activates nuclear Abl tyrosine kinase in wild type and *Rb*-knockout fibroblasts, Abl activation is not observed in *Rb-MI* fibroblasts.[49] These results are consistent with the previous finding that RB binds to and inhibits the nuclear Abl kinase.[50] Thus, RB cleavage/degradation is a prerequisite for Abl activation in response to TNF/CHX.

The observation that TNF/CHX can still activate Abl in *Rb*-knockout cells suggests that RB loss is necessary but not sufficient to fully stimulate Abl kinase activity.[45] Given the fact that Abl can adopt an auto-inhibitory conformation,[5] it may not be surprising that its dissociation from RB is but the first step towards the full activation of its catalytic function.[5,51] The mechanism by which TNF activates nuclear Abl kinase downstream of RB degradation is presently unknown.

Cleavage of Abl Protein by Caspase

The Abl protein also contains several caspase cleavage sites that are distributed in the large C-terminal region (Fig. 1). Cleavage at the most C-terminal of these sites eliminates the nuclear export signal. As a result, cleavage of nuclear Abl at this site would prevent its export leading to its nuclear accumulation.[52] As discussed below, the pro-apoptotic function of Abl only manifested when it is activated in the nucleus. Thus, caspase cleavage-mediated nuclear accumulation of Abl could enhance the apoptotic response.

While cleavage of the C-terminal region can alter the subcellular distribution of Abl, it does not lead to kinase activation.[52] As discussed above, activation of nuclear Abl requires the cleavage/degradation of RB. RB binds to the kinase domain of Abl to prevent Abl activation.[50] Therefore, caspase cleavage of Abl C-terminal region is not likely to disrupt the inhibitory effect of RB. Whether Abl cleavage by caspase is required for the activation of its kinase by TNF/CHX remains to be determined.

Abl Contributes to TNF-Induced Apoptosis

The suggestion that Abl plays a role in TNF-induced apoptosis came from the observation that imatinib (STI571, Gleevec), a chemical inhibitor of Abl kinase, attenuates TNF-induced cell death in human U937 myeloid leukemia cells.[46] Imatinib also inhibited TNF-induced apoptosis in embryo fibroblasts derived from *Rb*-knockout mice.[45] Further evidence for Abl involvement in TNF-induced apoptosis came from genetic studies. *Rb*-knockout fibroblasts undergo apoptosis when stimulated by TNF alone, without the need for CHX. *Rb/Abl*-double knockout fibroblasts showed reduced apoptotic response to TNF when compared with *Rb$^{-/-}$Abl$^{+/+}$* fibroblasts prepared from littermate embryos.[45] *Abl*-knockout fibroblasts display apoptotic response to TNF/CHX, which is greatly enhanced by the restitution of Abl expression.[45] Moreover, thymocytes explanted from *Abl*-knockout mice are more resistant to TNF-induced death than wild type thymocytes from literate mice.[45] Taken together, these results establish Abl as a death signal transducer in TNF-induced apoptosis.

It should be noted that imatinib does not attenuate death induced by TNF/CHX in all cell types. Thus, TNF/CHX can induce Abl-dependent and Abl-independent cell death. The basis for the differential requirement of Abl in TNF-induced apoptosis is presently not understood. Interestingly, while *Abl*-knockout cells showed a reduced death response to TNF, their apoptotic response to FAS-ligand was comparable to that of wild type cells (Cho S, unpublished). Although FAS receptor and TNFRI both activate the initiator caspases to commence death signal transduction, the mechanism of initiator caspase activation by FAS is different from that by TNFRI.[53,54] The variable requirement for Abl in TNF-induced death in different cell lines, and the lack of involvement of Abl in FAS-mediated cell death indicate that Abl may not be a general transducer of death signal downstream of initiator caspases. Instead, Abl may function to amplify death signal from TNFRI but only under certain cell context.

Nucleo-Cytoplasmic Abl Shuttling Matters in Cell Death

As discussed above, apoptotic stimuli such as DNA damage or TNF activate the nuclear pool of Abl kinase. The Abl protein undergoes nucleo-cytoplasmic shuttling in proliferating cells,[55,56] but nuclear import of Abl is blocked in terminally differentiated myotubes[57] and in the malignant anaplastic thyroid cancer cells.[56] Inhibition of Abl nuclear import can be correlated with resistance to genotoxin-induced apoptosis in differentiated myocytes and in anaplastic thyroid cancer cells.[56,57] Interestingly, the oncogenic BCR-ABL tyrosine kinase, when trapped in the nucleus, can also stimulate cell death.[58] These observations have suggested that Abl kinase must be activated in the nucleus to stimulate apoptosis.[2]

The BCR-ABL protein does not undergo nuclear import despite the fact that it contains all three NLS from ABL (Fig. 1).[58] Therefore, the three NLS are masked or inactivated in the BCR-ABL protein. We have found that imatinib, when bound to the kinase domain of BCR-ABL, can cause the unmasking or activation of the NLS leading to nuclear import. However, imatinib does not cause nuclear accumulation of BCR-ABL because the NES drives nuclear export at a faster rate than import.[2] The NES of ABL interacts with Crm1/Exportin-1, which is sensitive to inactivation by leptomycin B (LMB), a bacterial metabolite that binds to Crm1/Exportin-1 with a high affinity and selectivity.[59] Treatment of cells with LMB causes the nuclear accumulation of ABL, but not BCR-ABL because BCR-ABL does not undergo nuclear import.[58] When cells are treated with imatinib and LMB, BCR-ABL accumulates in the nucleus.[58] The combined treatment with imatinib and LMB causes irreversible and complete killing of

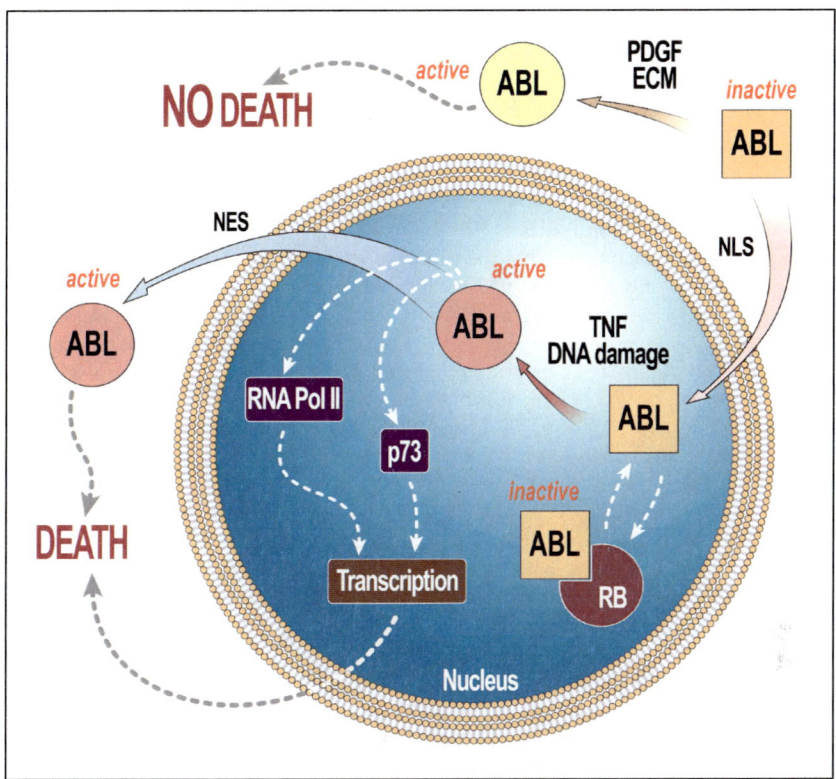

Figure 2. ABL subcellular location and the direction of its translocation are important in the regulation of cell death. Activation of cytoplasmic ABL by extracellular matrix (ECM) or growth factors (e.g., PDGF) is not sufficient to activate cell death. Because constitutively activated ABL mutants do not undergo nuclear import, we propose that activated cytoplasmic ABL is prohibited from entering the nucleus. The ABL NLSs may only be accessible to the nuclear import machinery when ABL adopts an inactivate conformation. The nuclear ABL is inactivated by its association with RB. Upon release from RB, either through RB phosphorylation or degradation, nuclear ABL can become activated by signals generated from damaged DNA or tumor necrosis factor (TNF) receptor. Activated nuclear ABL can stimulate cell death by transcription-dependent (through p73 and/or RNA polymerase II) mechanism. Activated nuclear ABL can exit to the cytoplasm and stimulate cell death by transcription-independent mechanism.

BCR-ABL-transformed cells, supporting the nuclear requirement for activated ABL kinase in inducing cell death.[58]

In recent unpublished studies, we have examined the pro-death function of Abl proteins that are either exclusively cytoplasmic (NLS-mutated), exclusively nuclear (Nuc),[56] or shuttling (wild type). These Abl proteins are activated by an inducible dimerization strategy.[56] When expressed in an appropriate cell context, dimerization of wild type or Abl-Nuk is sufficient to activate programmed cell death (XD Huang and JYJ Wang, unpublished). However, dimerization of cytoplasmic Abl does not initiate cell death (XD Huang and JYJ Wang, unpublished). This result demonstrates the requirement for Abl nuclear import in the induction of apoptosis. Interestingly, we found the wild type (shuttling) Abl is more efficient than the exclusively nuclear (Nuc) Abl in killing cells. Moreover, blocking the export of wild type Abl with LMB reduces its death-inducing activity (XD Huang and JYJ Wang, unpublished). Taken together, the current results suggest that nuclear import and export of Abl both contribute to cell death. However, the direction of Abl shutting matters: Abl must be activated in the nucleus to stimulate cell death, and nuclear exit of activated Abl can further enhance the death response (Fig. 2). This model would suggest that Abl may activate cell death effectors in the nucleus and the cytoplasm.

Death Effectors Regulated by Abl

Effectors That Regulate Transcription: p53, Mdm2, p73 and RNAPII

Abl and p53

Transcriptional regulation of death effectors is an important mechanism in the control of apoptosis. The best-known example is the transcriptional up-regulation of pro-apoptotic BH3 proteins, e.g., Puma and Noxa, by p53 in response to DNA damage. Indeed, IR-induced apoptosis in the developing central nervous system (CNS) is abrogated in *ATM*-knockout, *p53*-knockout or *Puma*-knockout mice.[40,60,61] The cumulative evidence strongly supports the induced expression of Puma to be an essential mechanism in IR-induced apoptosis. Activation of p53 by IR requires the ATM kinase, which not only phosphorylates p53 but also phosphorylates Mdm2, an inhibitor of p53.[62-64] Disruption of p53/Mdm2 complex by ATM-mediated phosphorylation events contributes to p53 activation and the induction of the expression of genes such as Puma and Noxa.

The functional interaction of Abl and p53 is suggested by several observations. Abl appears to phosphorylate Mdm2 and thus neutralizes its inhibitory effect on p53.[65,66] The human ABL protein might directly associate with p53 to enhance its transactivating function.[67] Excessive overproduction of Abl causes the inhibition of DNA replication in mouse fibroblasts, and this cytostatic effect of Abl is abrogated by the inactivation of p53, either through gene knockout or through the expression of HPV-E6 oncoprotein.[68,69] Thus, Abl is likely to activate p53 and causing p53-dependent G1-arrest.

However, activated Abl tyrosine kinase can stimulate death in *p53*-knockout fibroblasts, suggesting p53 is dispensable in Abl-initiated apoptosis.[56] Moreover, genotoxins can activate the endogenous Abl kinase to stimulate p53-independent cell death response.[70] As discussed above, we have found that *Abl*-knockout does not affect IR-induced CNS apoptosis, which is critically dependent on p53.[40] These results suggest the pro-apoptotic function of Abl and p53 can be unlinked. Of course, we cannot rule out the possibility that Abl may activate the apoptotic function of p53 in cell types other than embryo fibroblasts or neuroblasts.

Abl and p73

Interestingly, the induction of pro-apoptotic gene expression in the developing CNS not only requires p53, but also requires its related family members p63 and p73.[71] Ectopic expression of p73 can up-regulate Puma.[72] Therefore, the transcriptional regulation of death efforts appears to involve p53 and p73. The Abl tyrosine kinase may contribute to the transcriptional regulation of pro-apoptotic gene expression because it has been linked to the regulation of p53 and p73.

The interdependence between Abl and p73 in apoptosis activation has been observed in many cell types following exposure to genotoxins or to TNF. The Abl tyrosine kinase is required for DNA damage to activate p73 in fibroblasts and carcinoma cells.[26,35-37] DNA damage-induced stabilization of p73 protein is dependent on Abl.[26] DNA damage-induced acetylation of p73, mediated by the p300 acetyl-transferase, is also dependent on Abl.[73] Activated Abl tyrosine kinase does not induce cell death in *p73*-knockout fibroblasts, but its pro-apoptotic function can be restored through the reexpression of p73.[56] Ectopic expression of p73 does not induce apoptosis in *Abl*-null cells, but its pro-apoptotic function can be restored through the reexpression of Abl.[26] In anaplastic thyroid cancer cells that exclude Abl from the nucleus, genotoxins cannot activate the endogenous p73 protein to induce apoptosis.[56] However, this resistance phenotype can be overcome by the expression of Abl-Nuk, which is engineered to enter the nucleus constitutively.[56] Activation of nuclear Abl kinase by genotoxins in fibroblasts requires integrin signals.[70] Likewise, activation of p73 by genotoxins is also dependent on cell adhesion in these fibroblasts.[70] As discussed above, *Abl*-knockout thymocytes are resistant to TNF-induced apoptosis.[45] Interestingly, *p53*-knockout thymocytes remain sensitive to TNF, however, *p73*-knockout thymocytes share the phenotype of *Abl*-knockout thymocytes by being resistant to TNF-induced death.[45]

Taken together, the cumulative results have demonstrated a strong linkage between Abl and p73 in the activation of programmed cell death. The precise mechanism by which Abl causes the stabilization and activation of p73 has not been elucidated. Stimulation of p73-dependent transcription by Abl is likely to play an important role in DNA damage-induced apoptosis.[73] However, TNF-induced apoptosis does not require new gene expression. Therefore, how nuclear Abl and p73 contribute to the transcription-independent apoptotic response is presently unclear.[51]

Abl and RNA Polymerase II

In addition to the activation of p53 and p73, nuclear Abl tyrosine kinase can directly regulate transcription through the phosphorylation of RNA polymerase II (RNAPII). The catalytic subunit of eukaryotic RNAPII contains a C-terminal repeated domain (CTD) with the consensus sequence YSPTSPS repeated 52 times in the mammalian CTD. Activation of nuclear Abl tyrosine kinase by DNA damage leads to the tyrosine phosphorylation of the CTD[14,15,74] and this event can be associated with increased transcription.[13] Whether tyrosine phosphorylation of the RNAPII-CTD contributes to the pro-apoptotic function of Abl remains to be determined.

Abl and NF-kB

NF-κB protects cells from TNF-induced death.[75] Constitutive inhibition of NF-κB through the genetic modification of its inhibitor-IκB enhances cell death to TNFα.[53] Thus, TNF-induced apoptosis is inhibited rather than activated by transcription. The IκB inhibitor of NF-κB has been shown to interact with Abl and consequently acquire phosphorylation on its tyrosine 305 from Abl kinase activity.[76] Tyrosine phosphorylation to a degree stabilizes IκB, resulting in substantial accumulation of IκB protein in the nucleus to suppress NF-κB transcriptional activity. The abatement of NF-κB-dependent signaling by Abl exacerbates TNF-induced apoptosis,[76] which ties in well with previous results on the requirement of Abl in cell death by TNFα.[45]

Transcription-Independent Effectors: RAD9, Caspase-9 and Aph2

Abl and Nuclear Factors That Stimulate Cytochrome C Release

Current data suggest that nuclear Abl to act upstream of mitochondria, stimulating the release of apoptototic factors, such as cytochrome C[45] (Preyer M, Wang JYJ, unpublished). As discussed above, Abl can activate the transcription of BH3-only proteins through p53 and/or p73 to stimulate cytochrome C release. However, TNF can kill cells through transcription-independent mechanism. Because nuclear activation of Abl contributes to TNF-induced cell death, this suggests Abl may also activate apoptosis through transcription-independent mechanism.

Recent results have identified four nuclear proteins, which when translocated to the cytoplasm, can activate mitochondria-dependent apoptosis. These proteins are p53 itself,[77,78] TR3,[79,80] histone H1.2,[81] and Rad9.[82] As discussed above, Abl has been linked to the stabilization of p53.[65,66] It will be of interest to determine whether Abl can stimulate the cytoplasmic accumulation of p53 to regulate apoptosis through transcription-independent mechanism. We have found that Abl and p73 are required for TNF to kill cells.[45] At present, it is not known if p73 can stimulate transcription-independent apoptosis.[51] Because p73 also contains nuclear import and export signals,[83] it will be of interest to determine whether it also plays a cytoplasmic role in activating apoptosis. At present, there is no evidence linking Abl to TR3 or histone H1.2, although the latter has been shown to mediate in part DNA damage-induced apoptosis.[81]

The mammalian Rad9 protein, the ortholog of the fission yeast Rad9 gene product, contains a BH3-domain at its N-terminus.[82] It has been demonstrated that this Rad9-BH3 domain can interact with the anti-apoptotic Bcl-2 and Bcl-xL proteins, neutralizing their effects, and promoting mitochondrial release of cytochrome C.[82] The N-terminal BH3 domain of Rad9 appears to be released into the cytoplasm following caspase-dependent cleavage in staurosporine-treated cells.[84] It has been reported that Abl can interact with Rad9 following DNA damage, phosphorylating a tyrosine residue in the Rad9 BH3 domain and thereby stimulating the interaction between Rad9 and Bcl-2/Bcl-xL.[85] Despite these results, the knockout of Rad9 in mouse ES cells have been shown to sensitize rather than protect cells from genotoxin-induced death.[86,87] This phenotype is consistent with the fact that Rad9 is a component of the 9-1-1 complex, a DNA clamp that is specifically loaded unto damaged DNA and plays a critical role in DNA damage sensing and DNA repair.[88,89] As discussed above, DNA repair proteins can participate in DNA damage signal transduction. The mismatch repair proteins can correct replication errors but they can also stimulate cell death. Because Rad9 is a sensor of damaged DNA, it is formally possible that Rad9 may stimulate cell death through its SH3 domain under conditions of persistent damage. The pro-apoptotic function of Rad9 may be overshadowed by the pro-repair function of Rad9 in the knockout cells. The potential interaction between Abl and Rad9 in DNA damage-induced apoptosis is interesting but the relevance of this pathway awaits further demonstration.

Abl Interaction with Other Cytoplasmic Death Effectors

A recent report showed that Abl could phosphorylate caspase-9 in its pro-domain in human U937 cells or mouse fibroblasts upon exposure to cytosine arabinose (AraC), a nucleoside analog that inhibits DNA replication.[90] Tyrosine phosphorylation of caspase-9 appeared to enhance apoptotic response to IR or AraC, because ectopic expression of tysoine-mutated caspase-9 reduced cell death.[90] Abl also interacts through its C-terminus with Aph2 (Abl-philin 2), a resident of the endoplasmic reticulum (ER).[91] Aph2 is a novel protein with DHHC zinc-finger motif. Its over-expression mildly prompts cell death, an effect synergistically enhanced by the coexpression of Abl.[91] The lack of information on Aph2 at present makes it difficult to interpret this interesting observation.

Oncogenic BCR-ABL in Cell Death Regulation

Cytoplasmic BCR-ABL Promotes Cell Survival

The pro-death activity of Abl tyrosine kinase appears to be at odds with the finding that deregulated Abl kinase, such as Gag-Abl of Abelson murine leukemia virus and BCR-ABL of chromic myelogenous leukemia, have oncogenic functions. In particular, BCR-ABL tyrosine kinase has a strong pro-survival function that is well established by a large body of literature (Table 2). Expression of BCR-ABL in a variety of factor-dependent hematopoietic cell lines leads to the acquisition of factor independence.[92,93] For example, BCR-ABL obviates the requirement of interleukin 3 (IL-3) in murine pro-B Ba/F3, myeloid NFS/N1.H7 and 32Dcl3 lines,[92,94,95] and the requirement for GM-CSF in human Mo7e megakaryocytic cells.[93] BCR-ABL also induces VEGF, a pro-survival factor, in Ba/F3 cells.[96] High levels of osteopontin (OPN), which activates NF-κB in favor of cell survival,[97] are linked to BCR-ABL expression in 32D cells. The induction of survival factors, combined with the activation of Ras, PI3-kinase and Akt/PKB, Jak2 and Stat5 in BCR-ABL transformed cells establish a cellular environment that is highly resistant to a wide variety of pro-apoptotic stimuli, including DNA damage (Table 2).

BCR-ABL also suppresses the functions of pro-apoptotic proteins to afford further protection against cell death. Endogenous p53 protein level decreases from the persistent tyrosine kinase activity of BCR-ABL in FDCP-mix mouse born marrow cells.[98] The marked decrease in p53 protein expression may be explained by the concomitant increase in the expression of Mdm2 protein, the E3 ubiquitin ligase targeting p53 for degradation.[99] BCR-ABL, through a hitherto unknown mechanism, stimulates *mdm2* RNA translation to the effect of p53 down-regulation.[99] Besides up-regulating Bcl-2 and Bcl-xL anti-apoptotic proteins,[100-102]

Table 2. Effect of BCR-ABL on cell death induced by a variety of agents

Name	Categories	Experimental System	Ref.
Agents (Death protected by BCR-ABL)			
TNFα	inflamatory cytokine	HL-60/BCR-ABL	118
TRAIL	inflamatory cytokine	HL-60/BCR-ABL, K562	140
4-hydroperoxy -cyclophosp- hamide (4-HC)	alkylating agent	BaF3/BCR-ABL, FDC-P1/BCR-ABL	141
Anti-Fas Ab or FasL	Protagonistic Fas antibody and ligand	HL-60/BCR-ABL, K562/Fas	142,143
Ara-C	A Pyrimidine nucleoside analog inhibiting the synthesis of DNA	HL-60/BCR-ABL, FDCP-Mix A4/BCR-ABL, MO7e/BCR-ABL, K562	144-146
Camptothecin	topoisomerase I inhibitor	UT-7/BCR-ABL, HL-60/BCR-ABL, K562	145,147
Ceramide	sphingolipid	HL-60/BCR-ABL	144
Cisplatin (CDDP)	DNA covalent adduct formation and cross-linking	BaF3/BCR-ABL	148,149
Cytochrome C	activator of apoptosome	32D/BCR-ABL, K562, Rat-1 Fibroblast/BCR-ABL	104
Daunomycin (Daunorubicin)	DNA intercalation	BaF3/BCR-ABL, FDC-P1/BCR-ABL, MO7e/BCR-ABL	141,150
Etoposide	topoisomerase II inhibitor	BaF3/BCR-ABL, FDC-P1/BCR-ABL, HL-60/BCR-ABL,UT7/Bcr-Abl, K562	141,144, 147
Hydroxyurea	An inhibitor of ribonucleo- side diphosphate reductase	FDCP-Mix A4/BCR-ABL, MO7e/BCR-ABL, K562	146,150
Mitomycin C	DNA covalent adduct formation and crosslinking	BaF3/BCR-ABL	149
Mitoxantrone	DNA intercalation	K562	151
Nocodazole	promoter of tubulin depolymerization	HL-60/BCR-ABL	152
Paclitaxel	promoter of microtubule assembly	K562	153
Staurosporine	protein kinase inhibitor	HL-60/BCR-ABL	145
UV radiation	genotoxin	MO7e/BCR-ABL, 4A2+/BCR-ABL	154
Vincristine	Inhibitor of microtubule formation	HL-60/BCR-ABL	155
VM26 (teniposide)	topoisomerase II inhibitor	HL-60/BCR-ABL	100
γ-iradiation	genotoxin	BaF3/BCR-ABL, FDC-P1/BCR-ABL, 32D/BCR-ABL,CML CD34+ cells	156
Agents (Death NOT protected by BCR-ABL)			
Actinomycin D	inhibitor of transcription	HL-60/BCR-ABL, K562	157
Apicidin/ LAQ824/SAHA	HDAC inhibitors	HL-60/BCR-ABL, K562, LAMA-84, primary CML-BC cells	108-110, 158

Continued on next Page

Table 2. Continued

Name	Categories	Experimental System	Ref.
Arachidonic acid	prostaglandin	H7.bcr-abl A54, primary CML CD34+ cells	159,160
Arginine butyrate	butyric acid derivative	K562	161
As$_2$O$_3$	oxidative stress inducer	HL-60/BCR-ABL, MO7e/BCR-ABL, K562, primary CML CD34+ cells	105,106, 162
As$_4$S$_4$	oxidative stress inducer	K562, primary CML cells	163
Bortezomib/ Tripeptidyl aldehyde	proteasome inhibitors	KBM5/BCR-ABL, KBM7/BCR-ABL, Z-119/KBM5/BCR-ABL, K562, LAMA-84	111
Cepharanthine	biscocrourine alkaloid	K562	164
Cycloheximide	inhibitor of translation	HL-60/BCR-ABL, K562	165
Genasense	antisense oligonucleotide against Bcl-2	TF-1-R/BCR-ABL	115
Gerdanamycin/ 17-AAG	Hsp90 inhibitors	FDC-P2/BCR-ABL, HL-60/BCR-ABL, K562	107,166, 167
H$_2$O$_2$	oxidative stress inducer	BaF3/BCR-ABL	168
Hyperthermia	NA	Primary CML cells,	169,170
IFN-α	inflammatory cytokine	FDCP-Mix/BCR-ABL, MO7e/BCR-ABL, K562	171-173
Lyn siRNA	downregulation of Lyn kinase	K562, EM-2, EM-3, LAMA-84, primary CML-BC cells	116
Nitric oxide (NO)	oxidative stress inducer	K562, KCl22, KYO, LAMA-84, primary CML CD34+ cells	174
NK cells, LAK cells	cellular defense	K562	69,175
PBT-3	*Hepoxilin* analog	K562	176
Perillyl alcohol	monocyclic monoterpene	FDC-P1/BCR-ABL, 32D/BCR-ABL, K562	177,178
Pyrrolo-1,5-benzoxazepine	benzoxazepine derivative	K562	151
Rapamycin	mTOR inhibitor	BaF3/BCR-ABL	113
SCH66336	farnesyl transferase inhibitor	BaF3/BCR-ABL, primary CML cells	179-181
Telomestatin	telomerase inhibitor	K562, OM9;22, primary CML cells	182,180
Ubenimex/ Actinonin	aminopeptidase inhibitors	K562	184

Agents (Death sensitized by BCR-ABL)

TNF-α	inflamatory cytokine	BaF3/BCR-ABL	119
TRAIL	inflamatory cytokine	CML and Ph1(+)ALL derived cell lines	142
Ceramide	sphingolipid	BaF3/BCR-ABL	119

BCR-ABL prevents pro-apoptotic Bax and Bad proteins from translocating to mitochondria so as to abrogate potentially apoptotic responses.[103] Recently, BCR-ABL was shown to prevent caspase-9 activation despite the release of cytochrome C from the mitochondria.[104]

Nuclear BCR-ABL Stimulates Cell Death

As discussed above, the discrepancy between the pro-survival function of BCR-ABL kinase and the pro-apoptotic function of Abl kinase has been resolved. BCR-ABL is an exclusively cytoplasmic protein and does not undergo nuclear import.[2,58] From the cytoplasmic location, BCR-ABL stimulates mitogenic signal transducers to promote cell proliferation and cell survival. If BCR-ABL is allowed into the nucleus, it induces cell death and can cause the complete eradication of leukemic cells[58] (Minami Y, Wang JYJ, unpublished).

As mentioned above, our recent results have suggested that both nuclear and cytoplasmic Abl contribute to the activation of programmed cell death (Huang XD, Wang JYJ, unpublished). Induced dimerization of nuclear Abl is required to initiate the process, which can then be promoted by participation of the cytoplasmic Abl (Fig. 2) Induced dimerization of the cytoplasmic Abl, although contributing to cell death, cannot by itself initiate the death response (Fig. 2). These results have prompted us to consider the possibility that cytoplasmic BCR-ABL tyrosine kinase might also contribute to cell death under some death stimulating conditions. We therefore examined the literature for agents that can kill BCR-ABL transformed cells (Table 2).

Agents That Kill BCR-ABL Transformed Cells

It is clear that BCR-ABL-transformed cells are sensitive to killing by a number of agents (Table 2). Generally speaking, these agents can act through two different mechanisms. The more obvious mechanism of death is by agents that interfere with the pro-survival function of BCR-ABL. This can be achieved either through the direct inhibition of BCR-ABL expression/activity or by blocking a critical survival pathway activated by BCR-ABL. The best known example of such an agent is imatinib, which inhibits the BCR-ABL kinase to neutralize its pro-survival function. The second and less obvious mechanism is for an agent to kill BCR-ABL-transformed cells by initiating a death pathway that coopts BCR-ABL to facilitate cell death.

Inspection of the agents listed in Table 2 suggests many exert their toxic effects by the more obvious mechanism, i.e., to neutralize BCR-ABL. Among them are arsenic trioxide, which hinders BCR-ABL translation,[105,106] 17-allylamino 17-demethoxygeldanamycin (17-AAG), an antagonist of Hsp90, which shifts BCR-ABL from Hsp90 to Hsp70 and induces the proteasomal degradation of BCR-ABL.[107] Several other agents, e.g., inhibitors of histone deacetylase and proteasomes, also kill BCR-ABL transformed cells by down-regulating the expression of this oncoprotein.[108-112] On the other hand, rapamycin,[113,114] anti-sense oligonucleotide against Bcl-2[115] and siRNA against Lyn kinase[116] kill BCR-ABL transformed cells without affecting BCR-ABL expression. Rapamycin is an inhibitor of mTOR, a protein kinase activated by BCR-ABL and known to promote cell survival. Anti-sense oligonucleotides against Bcl-2 sensitizes a number of different types of leukemic cells to death, possibly through the destruction of Bcl-2, an important anti-apoptotic protein that maintains mitochondrial integrity. The Src-family of tyrosine kinases, such as Lyn, may contribute to the activation of BCR-ABL kinase.[117] Thus, knockdown of Lyn by siRNA may interfere with the pro-survival activity of BCR-ABL.[116]

At present, there is no direct evidence yet to support the hypothesis that some agents can kill BCR-ABL transformed cells by the second mechanism, i.e., by recruiting the cytoplasmic BCR-ABL as a collaborator in causing cytotoxicity. It is interesting to note that discrepant results have been obtained regarding the sensitivity of BCR-ABL-transformed cells to TNFα. With BCR-ABL expressing HL-60 cells, protection from TNFα-induced apoptosis has been reported[118] (Table 2). However, with BCR-ABL-transformed BaF3 cells, sensitization to TNFα- and ceramide-induced death has been reported[119] (Table 2). As discussed above, the Abl tyrosine kinase contributes to TNFα-induced death.[45] The observation that certain

BCR-ABL-transformed cells are more sensitive to death induced by TNFα suggests that cytoplasmic BCR-ABL may also have pro-death function that can be stimulated under the appropriate cell context, e.g., following the exposure to TNFα.

Conclusions and Future Prospects

The current evidence strongly supports a role for Abl tyrosine kinase in the regulation of cell death. In the developing embryos of *Rb* knockout mice, apoptosis in the emerging central nervous system and the embryonic liver can be rescued by the knockout of *Abl* (Borges H, Huntons IC, Wang JYJ, manuscript in preparation). Thus, Rb suppresses the apoptotic function of Abl during embryonic development. Activation of the Abl apoptotic function is controlled at several levels- its nuclear import, its interaction with inhibitors such as Rb and activators such as mismatch repair proteins and ATM. Nuclear Abl exerts its pro-apoptotic effects through the p53-family of transcription factors-p53 and p73. Whether Abl also interacts with and regulates the third member-p63 is presently unknown and needs to be investigated. Nuclear export of Abl appears to further enhance apoptosis, although activation of cytoplasmic Abl alone is not sufficient to induce cell death. Interestingly, the constitutively active BCR-ABL kinase, which is exclusively cytoplasmic, is a strong inhibitor of apoptosis. It is unclear if cytoplasmic Abl has anti-apoptotic function in mammalian cells. The worm Abl, presumably only present in the cytoplasm, has been shown to protect against ionizing radiation-induced death in germ cells. Many questions remain regarding the cell death regulatory function of Abl tyrosine kinase. The most challenging among them is how the death-regulatory function and the cytoskeleton-regulatory function of Abl are coordinated to regulate embryogenesis and tissue homeostasis throughout adult life.

References

1. Wang JY. Regulation of cell death by the Abl tyrosine kinase. Oncogene 2000; 19(49):5643-5650.
2. Zhu J, Wang JY. Death by Abl: A matter of location. Curr Top Dev Biol 2004; 59:165-192.
3. Woodring PJ, Hunter T, Wang JY. Regulation of F-actin-dependent processes by the Abl family of tyrosine kinases. J Cell Sci 2003; 116(Pt 13):2613-2626.
4. Nagar B, Hantschel O, Young MA et al. Structural basis for the autoinhibition of c-Abl tyrosine kinase. Cell 2003; 112(6):859-871.
5. Wang JY. Controlling Abl: Auto-inhibition and coinhibition? Nat Cell Biol 2004; 6(1):3-7.
6. Hantschel O, Superti-Furga G. Regulation of the c-Abl and Bcr-Abl tyrosine kinases. Nat Rev Mol Cell Biol 2004; 5(1):33-44.
7. Woodring PJ, Hunter T, Wang JY. Mitotic phosphorylation rescues Abl from F-actin-mediated inhibition. J Biol Chem 2005; 280(11):10318-10325.
8. Smith JM, Katz S, Mayer BJ. Activation of the Abl tyrosine kinase in vivo by Src homology 3 domains from the Src homology 2/Src homology 3 adaptor Nck. J Biol Chem 1999; 274(39):27956-27962.
9. Liu ZG, Baskaran R, Lea-Chou ET et al. Three distinct signalling responses by murine fibroblasts to genotoxic stress. Nature 1996; 384(6606):273-276.
10. Wang JY, Ki SW. Choosing between growth arrest and apoptosis through the retinoblastoma tumour suppressor protein, Abl and p73. Biochem Soc Trans 2001; 29(Pt 6):666-673.
11. Pendergast AM. The Abl family kinases: Mechanisms of regulation and signaling. Adv Cancer Res 2002; 85:51-100.
12. Kharbanda S, Yuan ZM, Weichselbaum R et al. Functional role for the c-Abl protein tyrosine kinase in the cellular response to genotoxic stress. Biochim Biophys Acta 1997; 1333(2):O1-7.
13. Baskaran R, Escobar SR, Wang JY. Nuclear c-Abl is a COOH-terminal repeated domain (CTD)-tyrosine (CTD)-tyrosine kinase-specific for the mammalian RNA polymerase II: Possible role in transcription elongation. Cell Growth Differ 1999; 10(6):387-396.
14. Baskaran R, Dahmus ME, Wang JY. Tyrosine phosphorylation of mammalian RNA polymerase II carboxyl-terminal domain. Proc Natl Acad Sci USA 1993; 90(23):11167-11171.
15. Baskaran R, Chiang GG, Wang JY. Identification of a binding site in c-Abl tyrosine kinase for the C-terminal repeated domain of RNA polymerase II. Mol Cell Biol 1996; 16(7):3361-3369.
16. Feller SM, Knudsen B, Hanafusa H. c-Abl kinase regulates the protein binding activity of c-Crk. Embo J 1994; 13(10):2341-2351.

17. Salgia R, Uemura N, Okuda K et al. CRKL links p210BCR/ABL with paxillin in chronic myelogenous leukemia cells. J Biol Chem 1995; 270(49):29145-29150.
18. Uemura N, Salgia R, Li JL et al. The BCR/ABL oncogene alters interaction of the adapter proteins CRKL and CRK with cellular proteins. Leukemia 1997; 11(3):376-385.
19. Woodring PJ, Litwack ED, O'Leary DD et al. Modulation of the F-actin cytoskeleton by c-Abl tyrosine kinase in cell spreading and neurite extension. J Cell Biol 2002; 156(5):879-892.
20. Kharbanda S, Ren R, Pandey P et al. Activation of the c-Abl tyrosine kinase in the stress response to DNA-damaging agents. Nature 1995; 376(6543):785-788.
21. Shafman T, Khanna KK, Kedar P et al. Interaction between ATM protein and c-Abl in response to DNA damage. Nature 1997; 387(6632):520-523.
22. Baskaran R, Wood LD, Whitaker LL et al. Ataxia telangiectasia mutant protein activates c-Abl tyrosine kinase in response to ionizing radiation. Nature 1997; 387(6632):516-519.
23. Shiloh Y. ATM and related protein kinases: Safeguarding genome integrity. Nat Rev Cancer 2003; 3(3):155-168.
24. Durocher D, Jackson SP. DNA-PK, ATM and ATR as sensors of DNA damage: Variations on a theme? Curr Opin Cell Biol 2001; 13(2):225-231.
25. Shiloh Y. ATM: Sounding the double-strand break alarm. Cold Spring Harb Symp Quant Biol 2000; 65:527-533.
26. Gong JG, Costanzo A, Yang HQ et al. The tyrosine kinase c-Abl regulates p73 in apoptotic response to cisplatin-induced DNA damage. Nature 1999; 399(6738):806-809.
27. Kolodner RD, Marsischky GT. Eukaryotic DNA mismatch repair. Curr Opin Genet Dev 1999; 9(1):89-96.
28. Flores-Rozas H, Clark D, Kolodner RD. Proliferating cell nuclear antigen and Msh2p-Msh6p interact to form an active mispair recognition complex. Nat Genet 2000; 26(3):375-378.
29. Drotschmann K, Hall MC, Shcherbakova PV et al. DNA binding properties of the yeast Msh2-Msh6 and Mlh1-Pms1 heterodimers. Biol Chem 2002; 383(6):969-975.
30. Lau PJ, Kolodner RD. Transfer of the MSH2. MSH6 complex from proliferating cell nuclear antigen to mispaired bases in DNA. J Biol Chem 2003; 278(1):14-17.
31. Bernstein C, Bernstein H, Payne CM et al. DNA repair/pro-apoptotic dual-role proteins in five major DNA repair pathways: Fail-safe protection against carcinogenesis. Mutat Res 2002; 511(2):145-178.
32. Sampath D, Rao VA, Plunkett W. Mechanisms of apoptosis induction by nucleoside analogs. Oncogene 2003; 22(56):9063-9074.
33. Achanta G, Pelicano H, Feng L et al. Interaction of p53 and DNA-PK in response to nucleoside analogues: Potential role as a sensor complex for DNA damage. Cancer Res 2001; 61(24):8723-8729.
34. Shangary S, Brown KD, Adamson AW et al. Regulation of DNA-dependent protein kinase activity by ionizing radiation-activated abl kinase is an ATM-dependent process. J Biol Chem 2000; 275(39):30163-30168.
35. Yuan ZM, Shioya H, Ishiko T et al. p73 is regulated by tyrosine kinase c-Abl in the apoptotic response to DNA damage. Nature 1999; 399(6738):814-817.
36. Agami R, Blandino G, Oren M et al. Interaction of c-Abl and p73alpha and their collaboration to induce apoptosis. Nature 1999; 399(6738):809-813.
37. White E, Prives C. DNA damage enables p73. Nature 1999; 399(6738):734-735, 737.
38. Huang Y, Yuan ZM, Ishiko T et al. Pro-apoptotic effect of the c-Abl tyrosine kinase in the cellular response to 1-beta-D-arabinofuranosylcytosine. Oncogene 1997; 15(16):1947-1952.
39. Yuan ZM, Huang Y, Ishiko T et al. Regulation of DNA damage-induced apoptosis by the c-Abl tyrosine kinase. Proc Natl Acad Sci USA 1997; 94(4):1437-1440.
40. Borges HL, Chao C, Xu Y et al. Radiation-induced apoptosis in developing mouse retina exhibits dose-dependent requirement for ATM phosphorylation of p53. Cell Death Differ 2004; 11(5):494-502.
41. Schumacher B, Hofmann K, Boulton S et al. The C. elegans homolog of the p53 tumor suppressor is required for DNA damage-induced apoptosis. Curr Biol 2001; 11(21):1722-1727.
42. Deng X, Hofmann ER, Villanueva A et al. Caenorhabditis elegans ABL-1 antagonizes p53-mediated germline apoptosis after ionizing irradiation. Nat Genet 2004; 36(8):906-912.
43. Yin XM. Bid, a critical mediator for apoptosis induced by the activation of Fas/TNF-R1 death receptors in hepatocytes. J Mol Med 2000; 78(4):203-211.
44. Yin XM. Signal transduction mediated by Bid, a pro-death Bcl-2 family proteins, connects the death receptor and mitochondria apoptosis pathways. Cell Res 2000; 10(3):161-167.
45. Chau BN, Chen TT, Wan YY et al. Tumor necrosis factor alpha-induced apoptosis requires p73 and c-ABL activation downstream of RB degradation. Mol Cell Biol 2004; 24(10):4438-4447.

46. Dan S, Naito M, Seimiya H et al. Activation of c-Abl tyrosine kinase requires caspase activation and is not involved in JNK/SAPK activation during apoptosis of human monocytic leukemia U937 cells. Oncogene 1999; 18(6):1277-1283.
47. Tan X, Wang JY. The caspase-RB connection in cell death. Trends Cell Biol 1998; 8(3):116-120.
48. Fattman CL, An B, Dou QP. Characterization of interior cleavage of retinoblastoma protein in apoptosis. J Cell Biochem 1997; 67(3):399-408.
49. Chau BN, Borges HL, Chen TT et al. Signal-dependent protection from apoptosis in mice expressing caspase-resistant Rb. Nat Cell Biol 2002; 4(10):757-765.
50. Welch PJ, Wang JY. A C-terminal protein-binding domain in the retinoblastoma protein regulates nuclear c-Abl tyrosine kinase in the cell cycle. Cell 1993; 75(4):779-790.
51. Wang JY. Nucleo-cytoplasmic communication in apoptotic response to genotoxic and inflammatory stress. Cell Res 2005; 15(1):43-48.
52. Barila D, Rufini A, Condo I et al. Caspase-dependent cleavage of c-Abl contributes to apoptosis. Mol Cell Biol 2003; 23(8):2790-2799.
53. Micheau O, Tschopp J. Induction of TNF receptor I-mediated apoptosis via two sequential signaling complexes. Cell 2003; 114(2):181-190.
54. Muppidi JR, Tschopp J, Siegel RM. Life and death decisions: Secondary complexes and lipid rafts in TNF receptor family signal transduction. Immunity 2004; 21(4):461-465.
55. Taagepera S, McDonald D, Loeb JE et al. Nuclear-cytoplasmic shuttling of C-ABL tyrosine kinase. Proc Natl Acad Sci USA 1998; 95(13):7457-7462.
56. Vella V, Zhu J, Frasca F et al. Exclusion of c-Abl from the nucleus restrains the p73 tumor suppression function. J Biol Chem 2003; 278(27):25151-25157.
57. Puri PL, Bhakta K, Wood LD et al. A myogenic differentiation checkpoint activated by genotoxic stress. Nat Genet 2002; 32(4):585-593.
58. Vigneri P, Wang JY. Induction of apoptosis in chronic myelogenous leukemia cells through nuclear entrapment of BCR-ABL tyrosine kinase. Nat Med 2001; 7(2):228-234.
59. Kudo N, Matsumori N, Taoka H et al. Leptomycin B inactivates CRM1/exportin 1 by covalent modification at a cysteine residue in the central conserved region. Proc Natl Acad Sci USA 1999; 96(16):9112-9117.
60. Herzog KH, Chong MJ, Kapsetaki M et al. Requirement for Atm in ionizing radiation-induced cell death in the developing central nervous system. Science 1998; 280(5366):1089-1091.
61. Villunger A, Michalak EM, Coultas L et al. p53- and drug-induced apoptotic responses mediated by BH3-only proteins puma and noxa. Science 2003; 302(5647):1036-1038.
62. Pandita TK, Lieberman HB, Lim DS et al. Ionizing radiation activates the ATM kinase throughout the cell cycle. Oncogene 2000; 19(11):1386-1391.
63. Khanna KK, Keating KE, Kozlov S et al. ATM associates with and phosphorylates p53: Mapping the region of interaction. Nat Genet 1998; 20(4):398-400.
64. Khosravi R, Maya R, Gottlieb T et al. Rapid ATM-dependent phosphorylation of MDM2 precedes p53 accumulation in response to DNA damage. Proc Natl Acad Sci USA 1999; 96(26):14973-14977.
65. Goldberg Z, Vogt Sionov R, Berger M et al. Tyrosine phosphorylation of Mdm2 by c-Abl: Implications for p53 regulation. Embo J 2002; 21(14):3715-3727.
66. Sionov RV, Moallem E, Berger M et al. c-Abl neutralizes the inhibitory effect of Mdm2 on p53. J Biol Chem 1999; 274(13):8371-8374.
67. Goga A, Liu X, Hambuch TM et al. p53 dependent growth suppression by the c-Abl nuclear tyrosine kinase. Oncogene 1995; 11(4):791-799.
68. Sawyers CL, McLaughlin J, Goga A et al. The nuclear tyrosine kinase c-Abl negatively regulates cell growth. Cell 1994; 77(1):121-131.
69. Roger R, Issaad C, Pallardy M et al. BCR-ABL does not prevent apoptotic death induced by human natural killer or lymphokine-activated killer cells. Blood 1996; 87(3):1113-1122.
70. Truong T, Sun G, Doorly M et al. Modulation of DNA damage-induced apoptosis by cell adhesion is independently mediated by p53 and c-Abl. Proc Natl Acad Sci USA 2003; 100(18):10281-10286.
71. Flores ER, Tsai KY, Crowley D et al. p63 and p73 are required for p53-dependent apoptosis in response to DNA damage. Nature 2002; 416(6880):560-564.
72. Melino G, Bernassola F, Ranalli M et al. p73 Induces apoptosis via PUMA transactivation and Bax mitochondrial translocation. J Biol Chem 2004; 279(9):8076-8083.
73. Costanzo A, Merlo P, Pediconi N et al. DNA damage-dependent acetylation of p73 dictates the selective activation of apoptotic target genes. Mol Cell 2002; 9(1):175-186.
74. Baskaran R, Chiang GG, Mysliwiec T et al. Tyrosine phosphorylation of RNA polymerase II carboxyl-terminal domain by the Abl-related gene product. J Biol Chem 1997; 272(30):18905-18909.

75. Beg AA, Baltimore D. An essential role for NF-kappaB in preventing TNF-alpha-induced cell death. Science 1996; 274(5288):782-784.
76. Kawai H, Nie L, Yuan ZM. Inactivation of NF-kappaB-dependent cell survival, a novel mechanism for the proapoptotic function of c-Abl. Mol Cell Biol 2002; 22(17):6079-6088.
77. Mihara M, Erster S, Zaika A et al. p53 has a direct apoptogenic role at the mitochondria. Mol Cell 2003; 11(3):577-590.
78. Schuler M, Maurer U, Goldstein JC et al. p53 triggers apoptosis in oncogene-expressing fibroblasts by the induction of Noxa and mitochondrial Bax translocation. Cell Death Differ 2003; 10(4):451-460.
79. Li H, Kolluri SK, Gu J et al. Cytochrome c release and apoptosis induced by mitochondrial targeting of nuclear orphan receptor TR3. Science 2000; 289(5482):1159-1164.
80. Jeong JH, Park JS, Moon B et al. Orphan nuclear receptor Nur77 translocates to mitochondria in the early phase of apoptosis induced by synthetic chenodeoxycholic acid derivatives in human stomach cancer cell line SNU-1. Ann NY Acad Sci 2003; 1010:171-177.
81. Konishi A, Shimizu S, Hirota J et al. Involvement of histone H1.2 in apoptosis induced by DNA double-strand breaks. Cell 2003; 114(6):673-688.
82. Komatsu K, Miyashita T, Hang H et al. Human homologue of S. pombe Rad9 interacts with BCL-2/BCL-xL and promotes apoptosis. Nat Cell Biol 2000; 2(1):1-6.
83. Inoue T, Stuart J, Leno R et al. Nuclear import and export signals in control of the p53-related protein p73. J Biol Chem 2002; 277(17):15053-15060.
84. Lee MW, Hirai I, Wang HG. Caspase-3-mediated cleavage of Rad9 during apoptosis. Oncogene 2003; 22(41):6340-6346.
85. Yoshida K, Komatsu K, Wang HG et al. c-Abl tyrosine kinase regulates the human Rad9 checkpoint protein in response to DNA damage. Mol Cell Biol 2002; 22(10):3292-3300.
86. Hopkins KM, Auerbach W, Wang XY et al. Deletion of mouse rad9 causes abnormal cellular responses to DNA damage, genomic instability, and embryonic lethality. Mol Cell Biol 2004; 24(16):7235-7248.
87. Loegering D, Arlander SJ, Hackbarth J et al. Rad9 protects cells from topoisomerase poison-induced cell death. J Biol Chem 2004; 279(18):18641-18647.
88. Parrilla-Castellar ER, Arlander SJ, Karnitz L. Dial 9-1-1 for DNA damage: The Rad9-Hus1-Rad1 (9-1-1) clamp complex. DNA Repair (Amst) 2004; 3(8-9):1009-1014.
89. Wang JY, Cho SK. Coordination of repair, checkpoint, and cell death responses to DNA damage. Adv Protein Chem 2004; 69:101-135.
90. Raina D, Pandey P, Ahmad R et al. c-Abl tyrosine kinase regulates caspase-9 autocleavage in the apoptotic response to DNA damage. J Biol Chem 2005; 280(12):11147-11151.
91. Li B, Cong F, Tan CP et al. Aph2, a protein with a zf-DHHC motif, interacts with c-Abl and has pro-apoptotic activity. J Biol Chem 2002; 277(32):28870-28876.
92. Daley GQ, Baltimore D. Transformation of an interleukin 3-dependent hematopoietic cell line by the chronic myelogenous leukemia-specific P210bcr/abl protein. Proc Natl Acad Sci USA 1988; 85(23):9312-9316.
93. Sirard C, Laneuville P, Dick JE. Expression of bcr-abl abrogates factor-dependent growth of human hematopoietic M07E cells by an autocrine mechanism. Blood 1994; 83(6):1575-1585.
94. Mandanas RA, Boswell HS, Lu L et al. BCR/ABL confers growth factor independence upon a murine myeloid cell line. Leukemia 1992; 6(8):796-800.
95. Matulonis U, Salgia R, Okuda K et al. Interleukin-3 and p210 BCR/ABL activate both unique and overlapping pathways of signal transduction in a factor-dependent myeloid cell line. Exp Hematol 1993; 21(11):1460-1466.
96. Mayerhofer M, Valent P, Sperr WR et al. BCR/ABL induces expression of vascular endothelial growth factor and its transcriptional activator, hypoxia inducible factor-1alpha, through a pathway involving phosphoinositide 3-kinase and the mammalian target of rapamycin. Blood 2002; 100(10):3767-3775.
97. Vejda S, Piwocka K, McKenna SL et al. Autocrine secretion of osteopontin results in degradation of I kappa B in Bcr-Abl-expressing cells. Br J Haematol 2005; 128(5):711-721.
98. Pierce A, Spooncer E, Wooley S et al. Bcr-Abl protein tyrosine kinase activity induces a loss of p53 protein that mediates a delay in myeloid differentiation. Oncogene 2000; 19(48):5487-5497.
99. Trotta R, Vignudelli T, Candini O et al. BCR/ABL activates mdm2 mRNA translation via the La antigen. Cancer Cell 2003; 3(2):145-160.
100. Amarante-Mendes GP, Finucane DM, Martin SJ et al. Anti-apoptotic oncogenes prevent caspase-dependent and independent commitment for cell death. Cell Death Differ 1998; 5(4):298-306.

101. Sanchez-Garcia I, Grutz G. Tumorigenic activity of the BCR-ABL oncogenes is mediated by BCL2. Proc Natl Acad Sci USA 1995; 92(12):5287-5291.
102. Uckun FM, Yang Z, Sather H et al. Cellular expression of antiapoptotic BCL-2 oncoprotein in newly diagnosed childhood acute lymphoblastic leukemia: A Children's Cancer Group Study. Blood 1997; 89(10):3769-3777.
103. Keeshan K, Cotter TG, McKenna SL. High Bcr-Abl expression prevents the translocation of Bax and Bad to the mitochondrion. Leukemia 2002; 16(9):1725-1734.
104. Deming PB, Schafer ZT, Tashker JS et al. Bcr-Abl-mediated protection from apoptosis downstream of mitochondrial cytochrome c release. Mol Cell Biol 2004; 24(23):10289-10299.
105. La Rosee P, Johnson K, O'Dwyer ME et al. In vitro studies of the combination of imatinib mesylate (Gleevec) and arsenic trioxide (Trisenox) in chronic myelogenous leukemia. Exp Hematol 2002; 30(7):729-737.
106. Nimmanapalli R, Bali P, O'Bryan E et al. Arsenic trioxide inhibits translation of mRNA of bcr-abl, resulting in attenuation of Bcr-Abl levels and apoptosis of human leukemia cells. Cancer Res 2003; 63(22):7950-7958.
107. Nimmanapalli R, O'Bryan E, Bhalla K. Geldanamycin and its analogue 17- allylamino- 17-demethoxygeldanamycin lowers Bcr-Abl levels and induces apoptosis and differentiation of Bcr-Abl-positive human leukemic blasts. Cancer Res 2001; 61(5):1799-1804.
108. Cheong JW, Chong SY, Kim JY et al. Induction of apoptosis by apicidin, a histone deacetylase inhibitor, via the activation of mitochondria-dependent caspase cascades in human Bcr-Abl-positive leukemia cells. Clin Cancer Res 2003; 9(13):5018-5027.
109. Nimmanapalli R, Fuino L, Bali P et al. Histone deacetylase inhibitor LAQ824 both lowers expression and promotes proteasomal degradation of Bcr-Abl and induces apoptosis of imatinib mesylate-sensitive or -refractory chronic myelogenous leukemia-blast crisis cells. Cancer Res 2003; 63(16):5126-5135.
110. Kim JS, Jeung HK, Cheong JW et al. Apicidin potentiates the imatinib-induced apoptosis of Bcr-Abl-positive human leukaemia cells by enhancing the activation of mitochondria-dependent caspase cascades. Br J Haematol 2004; 124(2):166-178.
111. Dou QP, McGuire TF, Peng Y et al. Proteasome inhibition leads to significant reduction of Bcr-Abl expression and subsequent induction of apoptosis in K562 human chronic myelogenous leukemia cells. J Pharmacol Exp Ther 1999; 289(2):781-790.
112. Dai Y, Rahmani M, Pei XY et al. Bortezomib and flavopiridol interact synergistically to induce apoptosis in chronic myeloid leukemia cells resistant to imatinib mesylate through both Bcr/Abl-dependent and -independent mechanisms. Blood 2004; 104(2):509-518.
113. Mayerhofer M, Aichberger KJ, Florian S et al. Identification of mTOR as a novel bifunctional target in chronic myeloid leukemia: Dissection of growth-inhibitory and VEGF-suppressive effects of rapamycin in leukemic cells. Faseb J 2005.
114. Mohi MG, Boulton C, Gu TL et al. Combination of rapamycin and protein tyrosine kinase (PTK) inhibitors for the treatment of leukemias caused by oncogenic PTKs. Proc Natl Acad Sci USA 2004; 101(9):3130-3135.
115. Tauchi T, Sumi M, Nakajima A et al. BCL-2 antisense oligonucleotide genasense is active against imatinib-resistant BCR-ABL-positive cells. Clin Cancer Res 2003; 9(11):4267-4273.
116. Ptasznik A, Nakata Y, Kalota A et al. Short interfering RNA (siRNA) targeting the Lyn kinase induces apoptosis in primary, and drug-resistant, BCR-ABL1(+) leukemia cells. Nat Med 2004; 10(11):1187-1189.
117. Plattner R, Kadlec L, DeMali KA et al. c-Abl is activated by growth factors and Src family kinases and has a role in the cellular response to PDGF. Genes Dev 1999; 13(18):2400-2411.
118. Fang G, Kim CN, Perkins CL et al. CGP57148B (STI-571) induces differentiation and apoptosis and sensitizes Bcr-Abl-positive human leukemia cells to apoptosis due to antileukemic drugs. Blood 2000; 96(6):2246-2253.
119. Maguer-Satta V, Burl S, Liu L et al. BCR-ABL accelerates C2-ceramide-induced apoptosis. Oncogene 1998; 16(2):237-248.
120. Takao N, Mori R, Kato H et al. c-Abl tyrosine kinase is not essential for ataxia telangiectasia mutated functions in chromosomal maintenance. J Biol Chem 2000; 275(2):725-728.
121. Miller HL, Lee Y, Zhao J et al. Atm and c-Abl cooperate in the response to genotoxic stress during nervous system development. Brain Res Dev Brain Res 2003; 145(1):31-38.
122. Dierov J, Dierova R, Carroll M. BCR/ABL translocates to the nucleus and disrupts an ATR-dependent intra-S phase checkpoint. Cancer Cell 2004; 5(3):275-285.
123. Foray N, Marot D, Randrianarison V et al. Constitutive association of BRCA1 and c-Abl and its ATM-dependent disruption after irradiation. Mol Cell Biol 2002; 22(12):4020-4032.
124. Cong F, Tang J, Hwang BJ et al. Interaction between UV-damaged DNA binding activity proteins and the c-Abl tyrosine kinase. J Biol Chem 2002; 277(38):34870-34878.

125. Oelgeschlager T. Regulation of RNA polymerase II activity by CTD phosphorylation and cell cycle control. J Cell Physiol 2002; 190(2):160-169.
126. Kharbanda S, Pandey P, Jin S et al. Functional interaction between DNA-PK and c-Abl in response to DNA damage. Nature 1997; 386(6626):732-735.
127. Jin S, Kharbanda S, Mayer B et al. Binding of Ku and c-Abl at the kinase homology region of DNA-dependent protein kinase catalytic subunit. J Biol Chem 1997; 272(40):24763-24766.
128. Yuan SS, Chang HL, Lee EY. Ionizing radiation-induced Rad51 nuclear focus formation is cell cycle-regulated and defective in both ATM(-/-) and c-Abl(-/-) cells. Mutat Res 2003; 525(1-2):85-92.
129. Yuan ZM, Huang Y, Ishiko T et al. Regulation of Rad51 function by c-Abl in response to DNA damage. J Biol Chem 1998; 273(7):3799-3802.
130. Chen G, Yuan SS, Liu W et al. Radiation-induced assembly of Rad51 and Rad52 recombination complex requires ATM and c-Abl. J Biol Chem 1999; 274(18):12748-12752.
131. Kitao H, Yuan ZM. Regulation of ionizing radiation-induced Rad52 nuclear foci formation by c-Abl-mediated phosphorylation. J Biol Chem 2002; 277(50):48944-48948.
132. Welch PJ, Wang JY. Abrogation of retinoblastoma protein function by c-Abl through tyrosine kinase-dependent and -independent mechanisms. Mol Cell Biol 1995; 15(10):5542-5551.
133. Welch PJ, Wang JY. Disruption of retinoblastoma protein function by coexpression of its C pocket fragment. Genes Dev 1995; 9(1):31-46.
134. Wen ST, Jackson PK, Van Etten RA. The cytostatic function of c-Abl is controlled by multiple nuclear localization signals and requires the p53 and Rb tumor suppressor gene products. Embo J 1996; 15(7):1583-1595.
135. Cho JW, Chung J, Baek WK et al. RB-resistant Abl kinase induces delayed cell cycle progression and increases susceptibility to apoptosis upon cellular stresses through interaction with p53. Int J Oncol 2003; 22(6):1193-1199.
136. Yu D, Khan E, Khaleque MA et al. Phosphorylation of DNA topoisomerase 1 by the c-Abl tyrosine kinase confers camptothecin sensitivity. J Biol Chem 2004.
137. Yuan ZM, Huang Y, Fan MM et al. Genotoxic drugs induce interaction of the c-Abl tyrosine kinase and the tumor suppressor protein p53. J Biol Chem 1996; 271(43):26457-26460.
138. Cheng WH, von Kobbe C, Opresko PL et al. Werner syndrome protein phosphorylation by abl tyrosine kinase regulates its activity and distribution. Mol Cell Biol 2003; 23(18):6385-6395.
139. Wen ST, Van Etten RA. The PAG gene product, a stress-induced protein with antioxidant properties, is an Abl SH3-binding protein and a physiological inhibitor of c-Abl tyrosine kinase activity. Genes Dev 1997; 11(19):2456-2467.
140. Nimmanapalli R, Porosnicu M, Nguyen D et al. Cotreatment with STI-571 enhances tumor necrosis factor alpha-related apoptosis-inducing ligand (TRAIL or apo-2L)-induced apoptosis of Bcr-Abl-positive human acute leukemia cells. Clin Cancer Res 2001; 7(2):350-357.
141. Bedi A, Barber JP, Bedi GC et al. BCR-ABL-mediated inhibition of apoptosis with delay of G2/M transition after DNA damage: A mechanism of resistance to multiple anticancer agents. Blood 1995; 86(3):1148-1158.
142. Uno K, Inukai T, Kayagaki N et al. TNF-related apoptosis-inducing ligand (TRAIL) frequently induces apoptosis in Philadelphia chromosome-positive leukemia cells. Blood 2003; 101(9):3658-3667.
143. McGahon AJ, Nishioka WK, Martin SJ et al. Regulation of the Fas apoptotic cell death pathway by Abl. J Biol Chem 1995; 270(38):22625-22631.
144. Amarante-Mendes GP, Naekyung Kim C, Liu L et al. Bcr-Abl exerts its antiapoptotic effect against diverse apoptotic stimuli through blockage of mitochondrial release of cytochrome C and activation of caspase-3. Blood 1998; 91(5):1700-1705.
145. Amarante-Mendes GP, McGahon AJ, Nishioka WK et al. Bcl-2-independent Bcr-Abl-mediated resistance to apoptosis: Protection is correlated with up regulation of Bcl-xL. Oncogene 1998; 16(11):1383-1390.
146. Xenaki D, Pierce A, Underhill-Day N et al. Bcr-Abl-mediated molecular mechanism for apoptotic suppression in multipotent haemopoietic cells: A role for PKCbetaII. Cell Signal 2004; 16(2):145-156.
147. McGahon A, Bissonnette R, Schmitt M et al. BCR-ABL maintains resistance of chronic myelogenous leukemia cells to apoptotic cell death. Blood 1994; 83(5):1179-1187.
148. Slupianek A, Hoser G, Majsterek I et al. Fusion tyrosine kinases induce drug resistance by stimulation of homology-dependent recombination repair, prolongation of G(2)/M phase, and protection from apoptosis. Mol Cell Biol 2002; 22(12):4189-4201.
149. Slupianek A, Schmutte C, Tombline G et al. BCR/ABL regulates mammalian RecA homologs, resulting in drug resistance. Mol Cell 2001; 8(4):795-806.

150. Thiesing JT, Ohno-Jones S, Kolibaba KS et al. Efficacy of STI571, an abl tyrosine kinase inhibitor, in conjunction with other antileukemic agents against bcr-abl-positive cells. Blood 2000; 96(9):3195-3199.
151. Mc Gee MM, Campiani G, Ramunno A et al. Pyrrolo-1,5-benzoxazepines induce apoptosis in chronic myelogenous leukemia (CML) cells by bypassing the apoptotic suppressor bcr-abl. J Pharmacol Exp Ther 2001; 296(1):31-40.
152. Nishii K, Kabarowski JH, Gibbons DL et al. ts BCR-ABL kinase activation confers increased resistance to genotoxic damage via cell cycle block. Oncogene 1996; 13(10):2225-2234.
153. Jamieson L, Carpenter L, Biden TJ et al. Protein kinase Ciota activity is necessary for Bcr-Abl-mediated resistance to drug-induced apoptosis. J Biol Chem 1999; 274(7):3927-3930.
154. Canitrot Y, Falinski R, Louat T et al. p210 BCR/ABL kinase regulates nucleotide excision repair (NER) and resistance to UV radiation. Blood 2003; 102(7):2632-2637.
155. Belhoussine R, Morjani H, Gillet R et al. Two distinct modes of oncoprotein expression during apoptosis resistance in vincristine and daunorubicin multidrug-resistant HL60 cells. Adv Exp Med Biol 1999; 457:365-381.
156. Santucci MA, Anklesaria P, Laneuville P et al. Expression of p210 bcr/abl increases hematopoietic progenitor cell radiosensitivity. Int J Radiat Oncol Biol Phys 1993; 26(5):831-836.
157. Abdelhaleem M. The actinomycin D-induced apoptosis in BCR-ABL-positive K562 cells is associated with cytoplasmic translocation and cleavage of RNA helicase A. Anticancer Res 2003; 23(1A):485-490.
158. Xu Y, Voelter-Mahlknecht S, Mahlknecht U. The histone deacetylase inhibitor suberoylanilide hydroxamic acid down-regulates expression levels of Bcr-abl, c-Myc and HDAC3 in chronic myeloid leukemia cell lines. Int J Mol Med 2005; 15(1):169-172.
159. Rizzo MT, Pudlo N, Farrell L et al. Specificity of arachidonic acid-induced inhibition of growth and activation of c-jun kinases and p38 mitogen-activated protein kinase in hematopoietic cells. Prostaglandins Leukot Essent Fatty Acids 2002; 66(1):31-40.
160. Rizzo MT, Regazzi E, Garau D et al. Induction of apoptosis by arachidonic acid in chronic myeloid leukemia cells. Cancer Res 1999; 59(19):5047-5053.
161. Urbano A, Koc Y, Foss FM. Arginine butyrate downregulates p210 bcr-abl expression and induces apoptosis in chronic myelogenous leukemia cells. Leukemia 1998; 12(6):930-936.
162. Puccetti E, Guller S, Orleth A et al. BCR-ABL mediates arsenic trioxide-induced apoptosis independently of its aberrant kinase activity. Cancer Res 2000; 60(13):3409-3413.
163. Yin T, Wu YL, Sun HP et al. Combined effects of As4S4 and imatinib on chronic myeloid leukemia cells and BCR-ABL oncoprotein. Blood 2004; 104(13):4219-4225.
164. Mukai M, Che XF, Furukawa T et al. Reversal of the resistance to STI571 in human chronic myelogenous leukemia K562 cells. Cancer Sci 2003; 94(6):557-563.
165. Hood KA, West LM, Northcote PT et al. Induction of apoptosis by the marine sponge (Mycale) metabolites, mycalamide A and pateamine. Apoptosis 2001; 6(3):207-219.
166. Nimmanapalli R, O'Bryan E, Huang M et al. Molecular characterization and sensitivity of STI-571 (imatinib mesylate, Gleevec)-resistant, Bcr-Abl-positive, human acute leukemia cells to SRC kinase inhibitor PD180970 and 17-allylamino-17-demethoxygeldanamycin. Cancer Res 2002; 62(20):5761-5769.
167. Gorre ME, Ellwood-Yen K, Chiosis G et al. BCR-ABL point mutants isolated from patients with imatinib mesylate-resistant chronic myeloid leukemia remain sensitive to inhibitors of the BCR-ABL chaperone heat shock protein 90. Blood 2002; 100(8):3041-3044.
168. Maru Y, Bergmann E, Coin F et al. TFIIH functions are altered by the P210BCR-ABL oncoprotein produced on the Philadelphia chromosome. Mutat Res 2001; 483(1-2):83-88.
169. Jain SK, de Aos I, Inai Y et al. Inactivation of wild-type BCR/ABL tyrosine kinase in hematopoietic cells by mild hyperthermia. Leukemia 2000; 14(5):845-852.
170. Thijsen SF, van Oostveen JW, Schuurhuis GJ et al. Hypersensitivity of bcr-abl-positive progenitors to hyperthermia in patients with chronic myeloid leukemia. Leukemia 1997; 11(10):1762-1768.
171. Luchetti F, Gregorini A, Papa S et al. The K562 chronic myeloid leukemia cell line undergoes apoptosis in response to interferon-alpha. Haematologica 1998; 83(11):974-980.
172. Andrews IIIrd DF, Singer JW, Collins SJ. Effect of recombinant alpha-interferon on the expression of the bcr-abl fusion gene in human chronic myelogenous human leukemia cell lines. Cancer Res 1987; 47(24 Pt 1):6629-6632.
173. Grebenova D, Kuzelova K, Fuchs O et al. Interferon-alpha suppresses proliferation of chronic myelogenous leukemia cells K562 by extending cell cycle S-phase without inducing apoptosis. Blood Cells Mol Dis 2004; 32(1):262-269.

174. Ferry-Dumazet H, Mamani-Matsuda M, Dupouy M et al. Nitric oxide induces the apoptosis of human BCR-ABL-positive myeloid leukemia cells: Evidence for the chelation of intracellular iron. Leukemia 2002; 16(4):708-715.

175. Baron F, Turhan AG, Giron-Michel J et al. Leukemic target susceptibility to natural killer cytotoxicity: Relationship with BCR-ABL expression. Blood 2002; 99(6):2107-2113.

176. Qiao N, Lam J, Reynaud D et al. The hepoxilin analog PBT-3 induces apoptosis in BCR-ABL-positive K562 leukemia cells. Anticancer Res 2003; 23(5A):3617-3622.

177. Chen Y, Hu D. Effects of POH in combination with STI571 on the proliferation and apoptosis of K562 cells. J Huazhong Univ Sci Technolog Med Sci 2004; 24(1):41-44.

178. Clark SS, Perman SM, Sahin MB et al. Antileukemia activity of perillyl alcohol (POH): Uncoupling apoptosis from G0/G1 arrest suggests that the primary effect of POH on Bcr/Abl-transformed cells is to induce growth arrest. Leukemia 2002; 16(2):213-222.

179. Nakajima A, Tauchi T, Sumi M et al. Efficacy of SCH66336, a farnesyl transferase inhibitor, in conjunction with imatinib against BCR-ABL-positive cells. Mol Cancer Ther 2003; 2(3):219-224.

180. Peters DG, Hoover RR, Gerlach MJ et al. Activity of the farnesyl protein transferase inhibitor SCH66336 against BCR/ABL-induced murine leukemia and primary cells from patients with chronic myeloid leukemia. Blood 2001; 97(5):1404-1412.

181. Hoover RR, Mahon FX, Melo JV et al. Overcoming STI571 resistance with the farnesyl transferase inhibitor SCH66336. Blood 2002; 100(3):1068-1071.

182. Tauchi T, Shin-Ya K, Sashida G et al. Activity of a novel G-quadruplex-interactive telomerase inhibitor, telomestatin (SOT-095), against human leukemia cells: Involvement of ATM-dependent DNA damage response pathways. Oncogene 2003; 22(34):5338-5347.

183. Tauchi T, Nakajima A, Sashida G et al. Inhibition of human telomerase enhances the effect of the tyrosine kinase inhibitor, imatinib, in BCR-ABL-positive leukemia cells. Clin Cancer Res 2002; 8(11):3341-3347.

184. Fujisaki T, Otsuka T, Gondo H et al. Bestatin selectively suppresses the growth of leukemic stem/progenitor cells with BCR/ABL mRNA transcript in patients with chronic myelogeneous leukemia. Int Immunopharmacol 2003; 3(6):901-907.

CHAPTER 5

Regulation of Cytoskeletal Dynamics and Cell Morphogenesis by Abl Family Kinases

Anthony J. Koleske*

Abstract

Abelson (Abl) family nonreceptor tyrosine kinases are essential regulators of cell morphogenesis in developing metazoan organisms. Mutant animals that lack Abl kinases exhibit defects in epithelial and neuronal morphogenesis. In cultured cells, the vertebrate Abl and Abl-related gene (Arg) proteins promote formation of actin-based protrusive structures, such as filopodia and lamellipodia. Abl family kinases act as relays that coordinate changes in cytoskeletal structure in response growth factors and adhesion receptor activation. These cytoskeletal rearrangements are achieved through the ability of these kinases to control the Rho and Rac GTPases and to stimulate assembly of protein complexes that activate nucleation of actin filaments by the Arp2/3 complex. Abl and Arg also contain extended C-termini that bind directly to F-actin and microtubules and may mediate interactions between these cytoskeletal networks in cells. Arg, for instance, can promote the cooperative assembly of an F-actin-rich scaffold in cells, which may serve as a base for the elaboration of actin-rich protrusive structures in cells.

Introduction

Changes in cell shape and migration are powered by dynamic rearrangements of the actin cytoskeleton. These changes are coordinated in space and time by protein machines that regulate actin filament polymerization/depolymerization, organize actin filaments into bundles or networks, and push or pull on these actin superstructures. These actin-based machines are controlled by signals from cell surface receptors that provide continuous updates on the extracellular physicochemical environment.

Abelson (Abl) family nonreceptor tyrosine kinases are essential regulators of cytoskeletal rearrangements in developing metazoan organisms. Abl family kinases act as relays that coordinate changes in cytoskeletal structure in response to discrete external cues. In this chapter, I review the cellular and developmental processes regulated by Abl family kinases and discuss the molecular mechanisms by which Abl family kinases control these processes.

*Anthony J. Koleske—Department of Molecular Biophysics and Biochemistry, Department of Neurobiology, Interdepartmental Neuroscience Program, Yale University, SHMC-E31, 333 Cedar Street, New Haven, Connecticut 06520-8024, U.S.A. Email: anthony.koleske@yale.edu

Abl Family Kinases in Development and Disease, edited by Anthony Koleske.
©2006 Landes Bioscience and Springer Science+Business Media.

Abl Family Kinases: Basic Anatomy

The N-Terminal Half of Abl Family Kinases Is Highly Conserved

The Abl kinase family is composed of the Abl and Arg (Abl-related gene, aka Abl2) kinases in vertebrates,[1,2] the Abl kinase in flies,[3] and the Abl-1 kinase in worms.[4] Following a short variable region, Abl family kinases contain tandem Src homology (SH) 3, SH2, and tyrosine kinase domains (Fig. 1). The SH3 and SH2 domains regulate kinase activity. In the inactive state, the SH3 and SH2 domains form an inhibitory scaffold along the back face of the kinase domain that holds it in the "off" state.[5,6] Upon activation, Abl undergoes a conformational change in which the SH3 and SH2 are relieved from their inhibitory conformation and promote kinase activity by facilitating interactions with substrates.[7,8] The Chapter by Hantschel and Superti-Furga contains an excellent review of Abl kinase regulation.

The C-Terminal Halves of Abl Family Kinases Contain Actin and Microtubule-Binding Domains

Abl family kinases share less sequence conservation downstream of the tyrosine kinase domain. The functional features of the C-terminal half are most well-characterized in Abl and Arg. Immediately C-terminal to the kinase domain, Abl and Arg contain three conserved Pro-X-X-Pro (PXXP) motifs that serve as binding sites for SH3-domain-containing proteins[9-13] (Fig. 1). The C-terminal halves of Abl family kinases are unique among nonreceptor tyrosine kinases because they contain multiple domains that interact directly with actin and microtubules. Both Abl and Arg share C-terminal calponin homology F-actin-binding domains[14-16] (Fig. 1). A globular (G-) actin-binding domain precedes this F-actin-binding domain in Abl,[15] whereas in Arg, it is preceded by an I/LWEQ homology F-actin-binding domain and a microtubule-binding domain.[16,17] Based on sequence comparisons, the fly Abl and Abl-1 proteins in worms are likely to contain an F-actin binding domain at their C-termini, but the actin-binding properties of these family members have not yet been examined.

Abl Contains Nuclear Localization and Export Sequences

In addition to the features that Abl shares with Arg, the Abl C-terminal half also contains a DNA-binding domain with 3 HMG-like motifs,[18] 3 nuclear localization sequences,[19] and a nuclear export sequence[20] (Fig. 1). These features appear to be important for Abl to function in the cell nucleus. The Chapter by Wang, Minami, and Zhu contains an excellent review of Abl's nuclear functions.

The Bcr-Abl Oncoprotein

Mutant forms of Abl cause chronic myelogenous leukemia (CML) and acute lymphocytic leukemia (ALL) in humans. Oncogenic activation of Abl is most commonly associated with a chromosomal translocation that fuses sequences encoding a portion of the Bcr protein to an N-terminal part of Abl resulting in a hybrid gene encoding the Bcr-Abl oncoprotein.[21] A coiled-coil motif in Bcr promotes oligomerization of the Bcr-Abl fusion protein[22,23] (Fig. 1). This oligomerization promotes kinase autophosphorylation and activates Abl tyrosine kinase activity,[24-26] leading to the activation of anti-apoptotic and mitogenic signaling pathways.[27-29] Oligomerization also promotes binding of Bcr-Abl to F-actin stress fibers,[24] which may contribute to the alterations in cytoskeletal structure[30] and cell adhesion systems observed in Bcr-Abl-transformed cells.[31]

Roles for Abl Family Kinases in Cellular Morphogenesis

Genetic studies reveal that Abl family kinases are essential for normal epithelial and neuronal cell morphogenesis during development. Studies of Abl family kinase function in cultured cells support a model in which Abl family kinases relay information from adhesion receptors and growth factor receptors to direct specific changes in cytoskeletal structure and function.

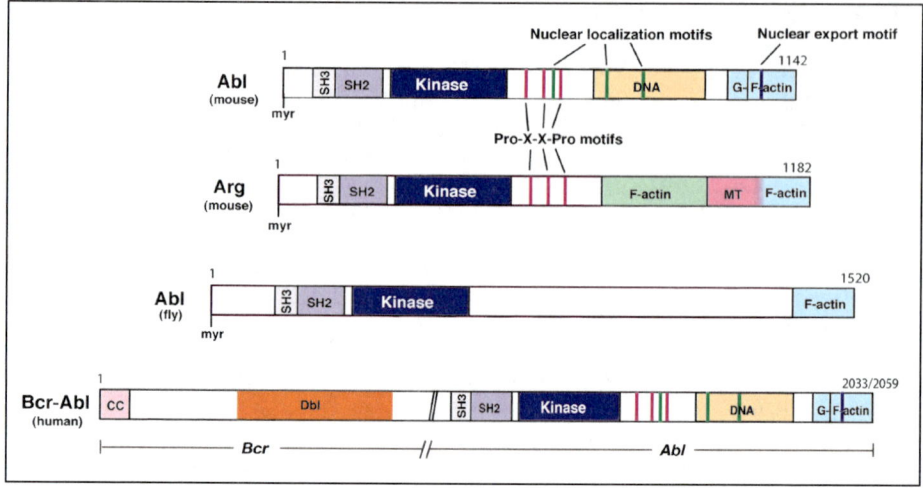

Figure 1. Domain stuctures of Abl, Arg, fly Abl, and Bcr-Abl. At least two alternatively-spliced amino termini have been reported for Abl and Arg.[116] Only the myristoylated (myr) form (type IV for Abl, type 1B for Arg) is shown here. Abl (mouse): Following the N-terminal exon, the indicated features are Src homology 3 (SH3), Src homology 2 (SH2), and tyrosine kinase (kinase) domains, 3 Pro-X-X-Pro motifs (pink stripes), 3 nuclear localization sequences (green stripes), a DNA binding domain, a globular- (G-) actin-binding domain, an F-actin binding domain (F-actin), and a nuclear export motif (purple stripe). Arg (mouse): Following the N-terminal exon, the indicated features are Src homology 3 (SH3), Src homology 2 (SH2), and tyrosine kinase (kinase) domains, 3 Pro-X-X-Pro motifs (pink stripes), an internal talin-like F-actin binding domain (light green F-actin), a microtubule-binding domain (MT) and a second calponin-homology F-actin-binding domain (light blue F-actin). Abl (fly): Based on sequence comparisons with mouse Abl and Arg, fly Abl is likely to be myristoylated at the N-terminus, but this has not been directly demonstrated. Like Abl and Arg, fly Abl contains the SH3, SH2, and kinase domains arranged in tandem. Sequence comparisons suggest that the fly Abl C-terminus contains an F-actin-binding domain, although this domain has not yet been shown to bind F-actin. Bcr-Abl (human): The N-terminal portion of Bcr is fused to Abl just N-terminal to the Abl SH3 domain. The Bcr moiety contains an N-terminal coiled-coil (CC) domain that mediates tetramerization of the fusion protein. Bcr also contains a Dbl domain with guanine nucleotide exchange activity for the Rho family GTPases Rho, Rac, and Cdc42. Different break points in Bcr can yield 210 kDa fusion proteins with different amino acid lengths.

This chapter will not discuss the regulation of neuronal morphogenesis and synaptic function by Abl family kinases, which is reviewed in an excellent chapter by Thompson and Van Vactor.

Abl and Arg Promote Cell Protrusions in Response to Growth Factors or Adhesive Cues

The first evidence that Abl family kinases regulate actin-based protrusions of the cell membrane came from observations of Bcr-Abl-transformed cells.[30] Salgia and colleagues showed that Bcr-Abl-transformed NIH3T3 fibroblasts exhibit more filopodia, lamellipodia, and membrane ruffles than their untransformed counterparts.[30] In contrast to the normal smooth round appearance of BaF3 hematopoietic cells, Bcr-Abl-expressing BaF3 cells extended numerous pseudopodia and were hypermotile on various adhesive surfaces.[30] Despite this clear demonstration that an activated form of Abl could promote changes in cytoskeletal structure, it was unclear how this activity related to the normal functions of Abl or Arg.

Subsequent studies of Abl- and/or Arg-deficient fibroblasts have clearly demonstrated that Abl and Arg act downstream of growth factor and adhesion receptors to promote cytoskeletal rearrangements. Application of platelet-derived growth factor (PDGF) to fibroblasts leads to dramatic actin-based ruffling of the cell periphery.[32,33] The finding that PDGF induces Abl kinase activity led Pendergast and colleagues to examine a role for Abl in membrane ruffling (Fig. 2A). Abl-deficient fibroblasts exhibit dramatically reduced membrane ruffling in response to PDGF, but this response can be restored by Abl reexpression.[33] These studies were the first to demonstrate that a normal function of Abl is to coordinate morphogenetic responses to a specific extracellular cue. Subsequent studies have further confirmed that Abl family kinases are required for membrane ruffling following cell exposure to PDGF and epithelial growth factor (EGF).[34] Although the function of these growth factor-induced ruffles is unknown, they may promote cell migration by helping cells sample three-dimensional space.

Abl family kinases also mediate cytoskeletal rearrangements in response to integrin-mediated adhesion to extracellular matrix proteins. Wild type fibroblasts are highly dynamic as they attach and spread on fibronectin-coated surfaces, frequently extending and retracting actin-based lamellipodial and filopodial protrusions.[17] Kymographic analysis reveals that Arg-deficient fibroblasts exhibit significantly fewer lamellipodial protrusions and retractions than wild type fibroblasts during adhesion and spreading on fibronectin (Fig. 2B). Reexpression of a functional Arg-yellow fluorescent protein (Arg-YFP) fusion restores normal lamellipodial behavior to these cells.[17] Importantly, Arg-YFP colocalizes with F-actin in these protrusive lamellipodial structures.[16,17] In a similar vein, *abl*[-/-]*arg*[-/-] fibroblasts, which have regular smooth peripheries as they adhere to and spread on fibronectin,[35] form multiple actin-based filopodial microspikes on fibronectin when Abl is reexpressed in these cells[35] (Fig. 2C). Abl localizes to the tips of these filopodia,[36] where it may regulate their protrusion, retraction, or adhesion to the extracellular matrix.

These studies demonstrate that Abl and Arg can each coordinate cell protrusions downstream of integrin adhesion receptors. The degree to which Abl and Arg overlap in the regulation of filopodia and lamellipodia is currently unclear. Simultaneous localization of fluorescently-labeled Abl and Arg in fibroblasts should reveal which protrusive structures contain which kinases. The analysis of *abl*[-/-]*arg*[-/-] cells reconstituted with Abl or Arg alone or both kinases should clarify which protrusions depend on which kinases.

Abl Family Kinases Regulate Epithelial Morphogenesis

Genetic studies in flies and mice have revealed a conserved function for Abl family kinases in the regulation of epithelial cell morphogenesis. The dorsal epidermis in flies is created through the convergent migration and joining of epithelial sheets, a process known as dorsal closure. Cells at the leading edge of the sheet uniformly elongate and assemble a regular actomyosin "purse-string" at their leading edges. The uniform constriction of this purse-string is believed to coordinate the drawing together of epithelial sheets. The Enabled (Ena) protein, a regulator of actin filament elongation, is enriched at the adherens junctions of these leading edge cells, where it likely helps coordinate the deposition and integrity of the actin purse-string.

Flies that lack both maternal and zygotic sources of fly Abl (*abl*[MZ] mutants) exhibit defects in dorsal epithelial closure.[37] *abl*[MZ] epithelial cells do not elongate uniformly perpendicular to the closure (Fig. 2D). Ena localization is less uniform in the *abl*[MZ] epithelial cells, with some cells containing larger concentrations of Ena than their neighbors. Cells with more Ena have more F-actin, leading to an uneven organization of the purse string. As a result of these defects, dorsal closure proceeds more slowly and irregularly in *abl*[MZ] embryos and in some embryos, holes in the epidermis remain. Consistent with a possible alteration in adherens junction structure and function, *abl*[MZ] mutants express reduced levels of α-catenin and β-catenin.

Formation of the neural tube, the precursor to the vertebrate brain and spinal cord has many similarities to dorsal closure in flies. The central nervous system begins as a flat sheet of ectoderm along the dorsomedial surface of the developing vertebrate embryo. The cells first undergo microtubule-dependent elongation to form the columnar epithelium of the neural plate.[38-40] Subsequent to this thickening, a contractile actomyosin latticework is formed at the

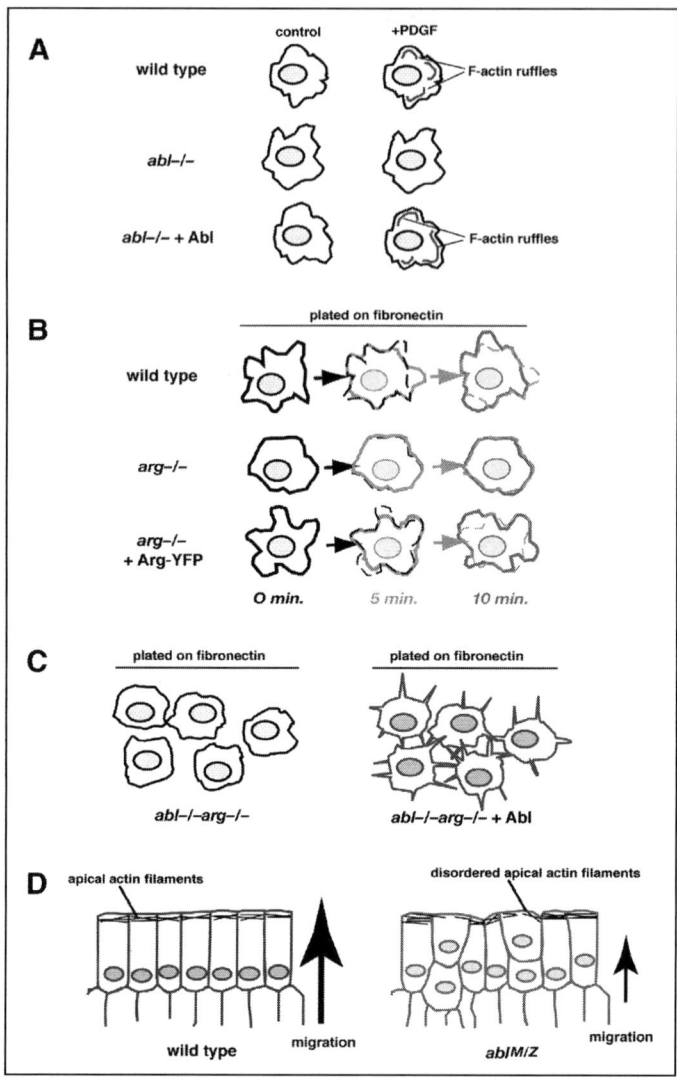

Figure 2. Abl family kinases are required for cellular morphogenesis. A) Top: treatment of wild type fibroblasts with PDGF induces multiple F-actin-rich ruffles. Middle: *abl*[-/-] fibroblasts treated with PDGF fail to exhibit ruffling. Bottom: reexpression of Abl in *abl*[-/-] fibroblasts (*abl*[-/-] + Abl) restores their ability to undergo PDGF-dependent F-actin ruffling. B) Top: wild type fibroblasts are highly dynamic as they spread on fibronectin with numerous lamellipodial protrusions and retractions. As a result, their shape is significantly different at 0 minutes (left), 5 minutes (middle) and 10 minutes (right). Middle: *arg*[-/-] fibroblasts are much less dynamic as they adhere and spread on fibronectin and their profiles change little over time.[17] Bottom: reexpression of an Arg-yellow fluorescent protein fusion in *arg*[-/-] cells (*arg*[-/-] + Arg-YFP) restores their dynamic behavior during adhesion and spreading on fibronectin. C) Left: *abl*[-/-]*arg*[-/-] fibroblasts exhibit smooth periphery as the cells spread on fibronectin. Right: reexpression of Abl in *abl*[-/-]*arg*[-/-] fibroblasts leads to a dramatic increase in filopodial microspikes as the cells spread on fibronectin.[35] D) Left: During dorsal closure in flies, epithelial cells at the dorsal leading edge elongate uniformly and migrate efficiently to closure point. Actin filaments form tightly ordered arrays at the apical surface of the leading edge cells (right). In *abl*[M/Z] mutant flies, which lack both maternal and zygotic sources of Abl, the cells do not elongate uniformly and migrate with reduced speed to the closure point.[37] The apical actin filaments are disordered in the leading edge cells of *abl*[M/Z] mutant epithelium.

apical surface of these neuroepithelial cells. Contraction of these filaments contributes to the polarized wedge shape that is essential for formation and maintenance of the neural tube.[38,40,41]

Abl and Arg are expressed in the mouse neuroepithelium, where they colocalize with the contractile apical actin latticework. $abl^{-/-}arg^{-/-}$ embryos suffer from severe defects in neural tube formation.[41a] Closure of the neural tube is delayed and even incomplete in some embryos. Upon closure, the neural tube of $abl^{-/-}arg^{-/-}$ embryos collapses into the central lumen of the neural tube. $abl^{-/-}arg^{-/-}$ neuroepithelial cells exhibit patchy disruptions of the apical actin latticework and ectopic actin-rich contractile structures at the basolateral surface. The fact that these defects are never observed in $abl^{-/-}$ or $arg^{-/-}$ single mutant embryos, suggests that Abl and Arg play redundant roles in the regulation of neuroepithelial morphogenesis.

Abl Regulates Cellularization in Fly Embryos

abl^{MZ} mutant flies have a high number of multinucleate cells.[42] This phenotype can be traced to defects in the formation of the pseudocleavage and cellularization furrows during the late syncytial stages of development.

In the late syncytial stages, nuclei localize to the embryo cortex where they continue through 4 coordinated rounds of cell division. During division, the nuclei become separated by an actin-rich pseudocleavage furrow that forms an inverted cup around the metaphase plate. After the final division, a cellularization furrow extends down and engulfs the entire nucleus with membrane.

The formation of pseudocleavage and cellularization furrows is controlled by dynamic rearrangements of the actin cytoskeleton. Actin forms a cap above interphase nuclei, but relocalizes and forms a pseudocleavage furrow during metaphase. Immunohistochemical staining showed that Abl is concentrated at apical junctions and its localization extends down into cleavage furrows during these events.[43] Some of these pseudocleavage furrows disintegrate in *abl* mutants, allowing two nuclei to share the same compartment. These defects correlate with increased staining for Ena, Diaphanous, the Arp2/3 complex, and F-actin at the apical surface and reduced F-actin in the pseudocleavage and cellularization furrows. A reduction in Ena levels suppresses the cellularization defects in *abl* mutants. Together, these studies suggest that Abl acts at the apical surface of the cell to manage Ena localization or activity.

Pathogens Exploit Abl-Regulated Pathways to Infect and Traffic within and between Cells

The ability of Abl family kinases to promote cell protrusions has been hijacked by some viruses and intracellular bacteria as a means to infect and move within and between cells.

The pathogenic bacterium *Shigella flexneri* infects nonphagocytic cells of the colonic mucosal lining. Contact between *Shigella* and a target cell induces actin-based protrusive structures that surround and eventually engulf the bacterium.[44] Abl and Arg localize to bacterial entry sites[45] and their kinase activities are induced by exposure to *Shigella*. This induction of kinase activity is critical: $abl^{-/-}arg^{-/-}$ cells or wild type cells treated with the Abl/Arg kinase inhibitor STI571 are resistant to *Shigella* infection. Activation of Abl/Arg by *Shigella* infection leads to increased phosphorylation of the Crk adaptor protein and is associated with downstream activation of the Rac and Cdc42 GTPases.[45] Activation of these downstream pathways is likely to underlie the complex actin-based rearrangements required for *Shigella* internalization.[45] In support of this, expression of a mutant Crk with a nonphosphorylatable substitution at the Abl/Arg phosphorylation site can inhibit *Shigella* infection. Once inside cells, *Shigella* escapes the vacuole and propels itself through the cytoplasm on a "comet tail" of polymerized actin. The IcsA coat protein on *Shigella* nucleates actin comet formation by recruiting and activating the N-WASp protein to promote Arp2/3 complex-dependent actin polymerization. It is not clear whether Abl family kinases are involved in the formation of the *Shigella* actin comet tail.

Vaccinia virus, a member of the poxvirus family, also utilizes cellular cytoskeletal systems to traffic within and between cells. Following replication and viral particle assembly in the cytoplasm, some particles become coated with membrane. These intracellular enveloped virions (IEVs) traffic to the cell surface along microtubules using the kinesin motor protein.[46,47] Once at the plasma membrane, the virus induces actin tail formation, which is believed to allow release of cell-associated enveloped virus (CEV) by propelling it away from the cell surface. Src family kinases localize to viral particles where they phosphorylate the vaccinia A36R protein, and recruit the Grb2 and Nck adaptors, N-WASP, and the Arp2/3 complex to promote formation of the actin comet tail.[48]

A recent study also shows that Abl and Arg localize to vaccinia virus comet tails.[49] Analysis of comet tail formation in mutant cell lines or in cells treated with different kinase inhibitors suggests that comet tail formation may require either Abl family or Src family kinases. Treatment with the selective Abl/Arg inhibitor STI571 (Gleevec) reduced the release of the enveloped virus.[49] Treatment of animals with STI571 also blocked the spread of vaccinia virus, although it remains possible that the inhibitor may be inhibiting Src or other kinases under the concentrations at which it was used. Nevertheless, this study introduces the possibility of using kinase inhibitors as adjuvants to vaccination to control poxvirus infection.

Mechanisms by Which Abl Family Kinases Regulate Cytoskeletal Structure

Abl family kinases respond to signals from cell surface receptors by directing changes in cytoskeletal structure. The past several years have seen a great increase in our understanding of the molecular mechanisms by which Abl family kinases act to control cytoskeletal rearrangements. A dominant theme is that Abl family kinases regulate the cytoskeletal structure by controlling the activity of the Rho family GTPases Rho and Rac. New studies also suggest that Abl family kinases can act directly on regulators of the Arp2/3 complex to promote actin polymerization. Finally, several recent experiments suggest that the C-terminus of Abl family kinases can act in a kinase-independent manner to regulate the structure of the F-actin and microtubule cytoskeletons.

A Primer: Rho Family GTPases Are Master Regulators of Cytoskeletal Rearrangements

Rho family GTPases, such as Rho, Rac, and Cdc42 act as molecular switches that regulate cytoskeletal rearrangements by cycling between an inactive GDP-bound form and an active GTP-bound form. In their active forms, Rho family GTPases interact with various effectors that regulate actin polymerization, F-actin severing, F-actin bundling, and actomyosin contractility.[50-52] Different Rho family GTPases interface with different effectors to produce different cytoskeletal structures, such as actin stress fibers, filopodia, or lamellipodia.[53-56]

Signals originating from cell surface receptors control the activity of Rho family GTPases by acting on two classes of regulatory molecules: guanine nucleotide exchange factors (GEFs) that activate Rho family GTPases by promoting the exchange of GTP for GDP and GTPase-activating proteins (GAPs) that inhibit Rho family GTPases by stimulating them to hydrolyze bound GTP. In some cases, the GEFs and GAPs are controlled directly by cell surface receptors. Alternatively, GEFs and GAPs can be controlled through the activity of a variety of protein kinases and cellular binding partners.

Regulation of the Rac GTPase by Abl Family Kinases

The Rac GTPase is a central regulator of lamellipodial formation and membrane ruffling in response to growth factor receptor signaling.[52,54] Active Rac promotes the recruitment and assembly of complexes containing WASp/WAVE-family proteins which activate the Arp2/3 complex to nucleate new actin filaments.[57]

Several observations in diverse experimental systems have implicated Abl family kinases in the regulation of Rac activity. Rac, Abl, and Arg all become activated in fibroblasts following PDGF treatment and both Rac and Abl activities are essential for PDGF-induced membrane ruffling.[33,58,59] Rac is also activated by Bcr-Abl and expression of a dominant-negative Rac can significantly delay disease onset in a mouse model of Bcr-Abl-induced leukemogenesis.[60] Similarly, dominant-negative Rac blocks some of the mitogenic effects of the v-Abl oncoprotein in fibroblasts.[61] Rac and Cdc42 are both activated upon cellular exposure to *Shigella* in wild type cells, but this activation is not observed in *abl⁻/⁻arg⁻/⁻* cells.[45] Thus, Abl family kinases control Rac activation in a wide variety of contexts.

Growth-Factor-Induced Rac Activation Requires Abl Phosphorylation of Sos-1

The murine Son-of-Sevenless 1 (Sos-1) can act as a GEF for both Ras and Rac, but acts as a Rac GEF when complexed with the Eps8 and Abi1 proteins following growth factor stimulation.[62] Exposure of cells to EGF or PDGF leads to increased phosphorylation of Sos-1[34] (Fig. 3). This growth factor-induced phosphorylation of Sos-1 is blocked in *abl⁻/⁻arg⁻/⁻* cells or in wild type cells treated with the Abl/Arg kinase inhibitor STI571. Sos-1 phosphorylation correlates with increased Rac GEF activity and this GEF activity is reduced by treatment with alkaline phosphatase. Growth factor-induced-Rac activation is reduced in *abl⁻/⁻arg⁻/⁻* cells, but can be increased upon reexpression of Abl. The reduced Rac activation has functional consequences, as *abl⁻/⁻* and *abl⁻/⁻arg⁻/⁻* cells exhibit a reduced ruffling response to PDGF.[33,34] These data demonstrate that Abl (and possibly Arg) contribute to Rac activation downstream of growth factor stimulation.

Interestingly, while Abl phosphorylation stimulates Sos-1 Rac GEF activity, its Ras GEF activity is high and is not affected by phosphorylation. The basis of this differential requirement for phosphorylation awaits further biochemical studies to elucidate the molecular mechanisms by which Abl phosphorylation regulates Sos-1.

Abl Modulates Rac Activation Downstream of Integrin-Mediated Adhesion

The CAS (Crk-associated substrate) protein becomes heavily phosphorylated in response to integrin-mediated adhesion (see refs. 63,64 and references therein). Tyrosine phosphorylation of CAS promotes its assembly into a Rac-activating complex that contains the adapter protein CrkII (Crk), Elmo, and the Rac GEF Dock180.[63,64] The Crk adapter contains an SH2 domain followed by two SH3 domains. Crk uses these domains to hold together the Rac activation complex: the Crk SH2 domain binds to phosphorylated CAS,[65] while the first Crk SH3 domain associates with Dock180.[66]

A number of observations suggest that Abl family kinases regulate the formation of the CAS/Crk/Dock180/Elmo complex. CrkII binds Abl and Arg and serves as their substrate in vitro and in vivo,[9,67,68] with Abl and Arg phosphorylating Crk at a single tyrosine residue (Y221) in a linker region between its two SH3 domains.[9,67] The Crk SH2 domain can bind to its own tyrosine-phosphorylated linker region, and therefore this phosphorylation has been proposed to inhibit Crk binding to other phosphotyrosine-containing proteins, including CAS.[67] Abl kinase activity levels do inversely correlate with Crk binding to CAS in vivo. For example, whereas *abl⁻/⁻arg⁻/⁻* fibroblasts have high levels of Crk:CAS containing-complexes, formation of these complexes is reduced in cells reexpressing Abl.[69]

It is difficult to reconcile a simple model in which Abl phosphorylation of Crk inhibits the formation of Crk:CAS complexes with the numerous examples (cited above) in which increased Abl kinase activity is associated with Rac activation. The more likely scenario is that phosphorylation of Crk represents an intermediate in the assembly of Crk/CAS/Elmo/Dock180 complexes (Fig. 3). Chodniewicz and Klemke propose a model in which Abl family kinases form complexes with Crk and help relocate it to the plasma membrane.[64] Once at the membrane, the Crk:Abl complex may be dissociated by phosphatases, leaving Crk available to associate with CAS and Dock180. Mutation of tyrosine 221 in Crk to phenylalanine promotes

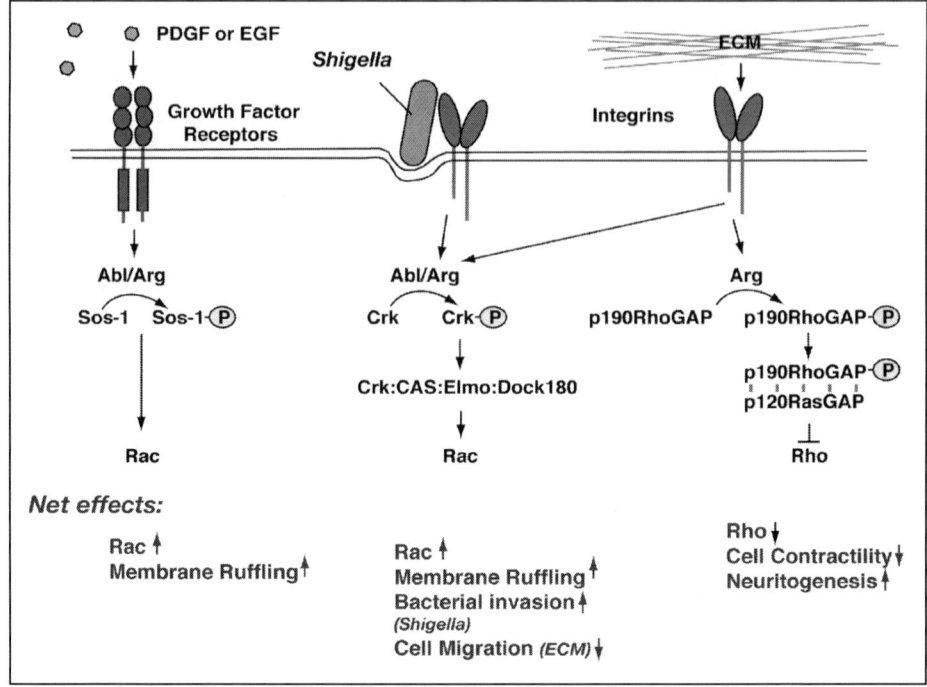

Figure 3. Abl and Arg regulate Rac and Rho downstream of diverse extracellular cues. Left) Binding of platelet-derived growth factor (PDGF) or epithelial growth factor (EGF) to their receptors activates Abl or Arg kinase activities.[3,59] Abl/Arg-mediated phosphorylation of Sos-1 activates Rac leading to increased membrane ruffling.[34] Middle) Binding of the *Shigella flexneri* bacterium to integrin receptors on the cell surfaces leads to activation of Abl/Arg kinase activity. Abl/Arg-directed phosphorylation of the Crk adaptor protein leads to the assembly of a Crk:CAS:Elmo:Dock180 complex that activates Rac,[64] leading to protrusive membrane ruffling and bacterial invasion. Right) Engagement of integrin receptors with the extracellular matrix (ECM) leads to activation of Abl and Arg kinase activity. Abl/Arg-mediated phosphorylation of Crk leads to activation of Rac (as outlined above). In this case, increased phosphorylation of Crk correlates with decreased overall cell motility.[69] Arg-mediated phosphorylation of p190RhoGAP promotes association with p120RasGAP, leading to decreased Rho activity and decreased cell contractility.[71] Arg-mediated inhibition of Rho in Neuro2A cells leads to increased neuritogenesis.[71]

increased complex formation with Abl[70] and blocks activation of Rac by Abl family kinases.[45] This residue may play an important role in dissociating Crk from Abl and promoting its association with CAS and Dock180.

Arg Is Required for Integrin-Dependent Inhibition of Rho

The RhoA (Rho) GTPase promotes the formation of actin stress fibers and focal adhesions in response to growth factor receptor signaling.[53] Active Rho activates Rho kinase, thereby promoting actomyosin contractility necessary for focal adhesion and stress fiber formation.[52]

Arg is particularly abundant in developing brain tissue and analysis of phosphotyrosine-containing proteins during postnatal brain development revealed that the 190 kD GTPase activating protein for Rho (p190RhoGAP) has reduced tyrosine phosphorylation in *arg*$^{-/-}$ brain. Arg phosphorylates p190RhoGAP directly on tyrosine 1105 (Y1105), thereby stimulating its ability to inhibit Rho.[71] Integrin-mediated adhesion to fibronectin promotes p190RhoGAP phosphorylation in wild type fibroblasts. This adhesion-dependent phosphorylation of p190RhoGAP is absent from *arg*$^{-/-}$ fibroblasts, but can be restored by reexpression of a functional

Arg-YFP fusion protein. Arg-deficient fibroblasts have larger, more numerous focal adhesions and larger stress fibers, a phenotype that appears to result from hyperactive Rho in these cells (A.L. Miller and A.J. Koleske, unpublished data). These observations suggest that Arg acts on p190RhoGAP to inhibit Rho following integrin-mediated cell attachment and spreading (Fig. 3).

Treatment of cultured neurons with the Abl/Arg inhibitor STI571 leads to simplification of neurite structure[35,72,73] that correlates with an increase in Rho activity.[72] Inhibition of Rho suppresses the effects of STI571 treatment on neurite structure.[72] This finding also supports a model in which Abl family kinases act to inhibit Rho.

Phosphorylation of Y1105 promotes formation of a complex between p190RhoGAP and the 120 kD GTPase activating protein for Ras (p120RasGAP).[74,75] This interaction is mediated in part by binding of one of two SH2 domains in p120RasGAP to the phosphorylated Y1105 in p190RhoGAP.[71,75] Increased p190RhoGAP phosphorylation and increased formation of the p190RhoGAP:p120RasGAP complex correlates with increased disassembly of actin stress fibers in Src-overexpressing or EGF-treated cells.[76,77] These data suggest that Arg-mediated phosphorylation activates p190RhoGAP by promoting its binding to p120RasGAP.

Previous studies had shown that p190RhoGAP is a major cellular target of Src family kinases.[78] Adhesion-dependent suppression of Rho activity does not occur in fibroblasts that lack the Src, Yes, and Fyn kinases (SYF fibroblasts).[79] An adhesion-dependent increase in p190RhoGAP phosphorylation is also not observed in these cells. Interestingly, Src family kinases phosphorylate p190RhoGAP at Y1105, the same site targeted by Arg.[75]

Arg and Src family kinases might represent alternative pathways leading from integrin receptors to p190RhoGAP phosphorylation. However, this model would not explain why elimination of either Arg or Src/Fyn/Yes eliminates all adhesion-dependent p190RhoGAP phosphorylation. A more likely scenario is that Arg and Src family kinases are in the same cascade, but only one of the kinases actually phosphorylates p190RhoGAP. Several studies have shown that Src is required for Abl kinase activation upon growth factor receptor stimulation.[33,80] Src family kinases activate Abl family kinases through phosphorylation of an activation loop tyrosine in the Abl/Arg kinase domain.[33,80-82] Arg phosphorylates p190RhoGAP in vitro with a K_M of 130 nM. Src can also phosphorylate p190RhoGAP in vitro, but it is unclear how its efficiency compares to that of Arg.[75] These observations suggest a model in which integrin engagement activates Src to phosphorylate Arg, which, in turn, phosphorylates and activates p190RhoGAP. Alternatively, Arg or Src might be required to set up an appropriate cellular microenvironment to allow phosphorylation of p190RhoGAP by the other kinase.

Abl Family Kinases Also Regulate Actin Structure Independently of Rho GTPases

In addition to their ability to regulate Rho and Rac, Abl family kinases interact with several actin regulatory complexes to control cytoskeletal rearrangements.

Abl Interactor (Abi) Proteins Mediate Interactions between Abl Family Kinases and Arp2/3 Regulatory Complexes

The Abi-1 and Abi-2 proteins were identified in two-hybrid screens for proteins that interact with Abl and Arg.[11-13] Abi-1 localizes to filopodial and lamellipodial tips, sites of active actin polymerization.[83] Abi-1 and Abi-2 have been identified in protein complexes containing WAVE proteins, which promote Arp2/3 complex-dependent actin polymerization. These complexes also contain the Nap1 and Pir120 proteins.[84-88] Together, these studies raise the possibility that Abi-1 and Abi-2 mediate interactions with Abl family kinases and WAVE proteins. Like Abi-1, Abl and Arg have been localized to sites of actin polymerization at lamellipodial protrusions and filopodial tips, and Abl was recently shown to be part of a complex containing WAVE2 and Abi-1.[89] Adhesion to fibronectin stimulates translocation of these proteins to the cell periphery where they colocalize with F-actin-rich lamellipodia. Activation of Abl kinase

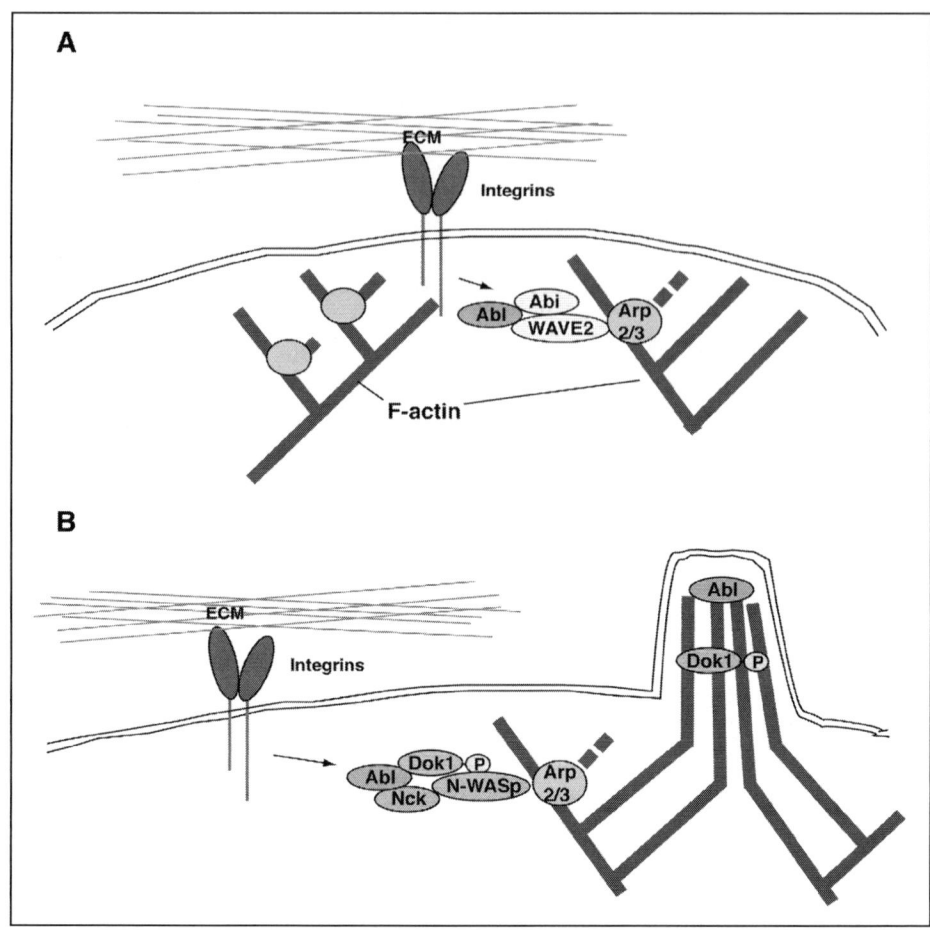

Figure 4. Abl regulates lamellipodial and filopodial formation via distinct mechanisms. A) Promotion of lamellipodial formation by Abl. Integrin-mediated adhesion to the extracellular matrix promotes the assembly of a protein complex containing Abl and Abi-1 and WAVE2.[89] This complex localizes to the lamellipodium periphery, where it activates the Arp2/3 complex to nucleate new actin filaments. B) Promotion of filopodial formation by Abl. Following activation by integrin-mediated adhesion, Abl phosphorylates Dok-1, allowing complex formation with the Nck-1 adaptor protein and N-WASp.[36] This protein complex activates actin filament nucleation and the resulting actin filaments become bundled and convergently extended into a filopodium. Abl localizes to the filopodial tips where it may regulate actin filament elongation.[36]

activity promotes tyrosine phosphorylation of Abi-1 and WAVE2. Mutation of a single tyrosine residue (Y150) in WAVE2 reduced tyrosine phosphorylation of WAVE2 and assembly of complexes containing WAVE2 and Abi-1. Abl also can stimulate the ability of purified Abi-1/WAVE2 to activate Arp2/3 complex-mediated actin polymerization in vitro. Together, these findings suggest that Abl mediates the formation of an actin polymerization regulatory complex containing in response to integrin-mediated adhesion (Fig. 4A). It is unclear at present whether the Abi-1:Abl:WAVE2 complexes also contain other proteins (e.g., PIR121, Nap1, HSPC300) that have been described in WAVE-containing complexes.

The Arg-interacting protein nArgBP2 was also recently shown to bind to WAVE1 and WAVE2.[90] nArgBP2 might serve as an alternative to Abi proteins to mediate interactions between Abl and WAVE family proteins.

Abl Phosphorylation of Dok Family Proteins Promotes Filopodia Formation

The Downstream of kinase (Dok1) protein was originally identified as a protein that was heavily phosphorylated in Bcr-Abl-transformed cells.[91,92] Subsequent studies have identified 5 family members (Dok1-5), each containing a pleckstrin homology (PH) domain, a phosphotyrosine binding (PTB) and a C-terminal tail with multiple potential tyrosine phosphorylation sites. Dok1 and Dok2 can serve as essential downstream effectors of Abl in filopodium/actin microspike formation.

Dok1 becomes tyrosine phosphorylated following integrin-mediated adhesion. This phosphorylated Dok1 binds avidly to the Abl SH2 domain.[36] Phosphorylation occurs on a single site (Y361) located in the C-terminal tail. In addition to having several potential phosphorylation sites, the Dok2 tail contains a PMMP motif that can interact selectively with the Abl SH3 domain.[93] Mutation of this motif abrogates Dok2 association with Abl when the two proteins are coexpressed. Abl expression also promotes Dok2 phosphorylation, which elevates Abl kinase activity, most likely by stabilizing interactions between the two proteins.

Expression of Dok-1 or Dok-2 with Abl leads to increased formation of F-actin microspikes.[36,93] Phosphorylation of Y361 in the Dok1 tail serves as a binding site for the Nck adaptor protein, which can bind to and activate cytoskeletal regulators such as WASp,[94] N-WASP,[95] and PAK.[96] These findings lead to a model in which Abl phosphorylation promotes the assembly of an Abl:Dok1:Nck complex that promotes actin polymerization into microspikes/filopodia (Fig. 4B). In support of this model, Abl and Dok1 both localize to filopodial tips. Cells that lack Dok-1 or both Nck1 and Nck2 form fewer actin-rich filopodia when plated on fibronectin. Ena/VASP family proteins are also found at the filopodial tips, where they may regulate the actin filament elongation.[97]

Abl Family Kinases Interact Functionally with Ena/VASP Proteins

Ena/VASP proteins, which include the fly Enabled (Ena) protein, Unc-34/Enabled in worms, and the mammalian Enabled (Mena), VASP (vasodilator-stimulated phosphoprotein), and EVL (Ena/VASP-like protein), regulate the elongation of actin filaments in a number of different cellular structures.[98] Despite the identification of the *enabled* gene over 15 years ago as a dosage-dependent modifier of the *abl* mutant fly phenotypes,[99] we still do not have a clear mechanistic picture of how Abl family kinases interact at a molecular level with Ena/VASP family proteins.

Genetic studies in flies suggest that Abl may regulate the localization of Ena/VASP family proteins. In the late syncytial divisions and during cellularization in wild type embryos, Ena is distributed at moderate levels at the apical cortex, with weaker levels throughout the cytoplasm.[42] In *abl* mutant flies, Ena is strongly enriched at apical surface in association with the increased F-actin structures. This observation suggests that Abl acts to inhibit Ena localization at the apical surface, possibly by interfering with the binding of Ena to one of its partners.

Ena/VASP proteins contain a central proline-rich domain (PRD) that can bind to the Abl SH3 domain. Binding of Abl to the PRD could prevent Ena from binding to other proteins via this domain. Moreover, phosphorylation of specific residues in the PRD of Ena, EVL, and VASP abolishes their ability to bind to SH3-domain-containing proteins.[100] These phosphorylations appear to regulate Ena/VASP protein interactions in response to discrete extracellular cues. For example, adhesion-dependent phosphorylation of VASP by protein kinase A prevents its binding to Abl during initial cell spreading.[117] Abl phosphorylation of Ena in vitro prevents its subsequent binding to the Abl SH3 domain. Vertebrate Abl and fly Abl can promote Mena and Ena phosphorylation respectively when they are coexpressed in cultured cells and this phosphorylation is enhanced by expression of Abi-1 or fly Abi, respectively.[101,102] These data suggest that Abi proteins might stabilize interactions between Ena/VASP proteins and Abl family kinases to allow for Ena/VASP phosphorylation. Future studies should clarify whether this event occurs as part of a response to physiological cues.

Abl Family Kinases Interact Directly with F-Actin and Microtubules

Abl family kinases are unique among nonreceptor tyrosine kinases in having extended C-terminal halves that contain domains that bind to actin, and, in the case of Arg, microtubules (see Fig. 1). In addition to their interactions with cytoskeletal regulatory proteins, Abl family kinases can influence cytoskeletal structure through these direct interactions with cytoskeletal components.

Arg Bundles F-Actin and May Form an F-Actin Scaffold to Recruit Cytoskeletal Regulators

Arg contains two distinct F-actin binding domains: a calponin homology (CH) F-actin binding domain at its C-terminus and an I/LWEQ (talin-like) F-actin binding domain located midway between its kinase domain and the CH domain.[16] The region in Abl corresponding to the CH domain in Arg can be separated into two distinct regions that bind to G- and F-actin.[15] An internal domain in Abl corresponding to the I/LWEQ domain can also bind weakly to F-actin (Maithreyi Krishnaswami and A.J.K. unpublished data).

Arg can assemble F-actin into tight bundles in vitro and this activity requires both the I/LWEQ and CH domains.[16] Reconstruction of electron microscopic images of short segments of Arg:F-actin complexes revealed that Arg could bind to F-actin in several modes.[103] Analysis of Arg mutants lacking either the I/LWEQ or CH domains revealed that the different modes resulted from binding via the different F-actin binding domains. The CH domain bound to subdomain 1 (SD1) of actin and induced a tilt in the actin filament. The I/LWEQ could bind in two modes to either SD1 or SD4. Importantly, only one mode of F-actin binding was found in each Arg:F-actin segment. These data showed that Arg cannot use both domains to bind simultaneously to the same filament. All of the binding modes leave the other domain available to bind an adjacent filament for bundle formation.

Arg binds cooperatively to F-actin in vitro during bundle formation.[16] This property may help concentrate Arg locally within cells, as Arg binding to F-actin would promote additional binding of Arg to F-actin locally. Electron microscopic structural studies of Arg bound to individual actin filaments show that Arg binding induces structural changes to the actin filament, changing both the structure of actin subunits and their helical pitch.[103] These structural changes are highly cooperative and propagate along patches of unoccupied actin filament. Importantly, some of the changes in actin filament structure resemble the F-actin conformation found in Arg:F-actin complexes. These findings suggest Arg-induced structural alterations to the actin filament could promote Arg binding to F-actin at sites distal to the initial binding site.

An important unresolved issue is whether Arg organizes F-actin into bundles or higher-order structures within cells. In fibroblasts attaching to fibronectin, Arg localizes to the periphery where it promotes the formation of F-actin-rich protrusive lamellipodial structures. Abl can also bundle F-actin in vitro, and is found in association with F-actin-rich structures in a variety of cell types.[36,104] An Arg C-terminal fragment (Arg688-1182) containing the two F-actin-binding domains is sufficient to form F-actin-rich structures at the cell periphery. These observations suggest that Arg can organize actin filaments into bundles or other higher order structures in vivo. However, while Arg induces highly dynamic F-actin-rich structures that protrude and retract, the F-actin-rich structures induced by Arg688-1182 are largely static. These findings suggest that Arg uses its 688-1182 fragment to localize to the periphery and locally organize F-actin structure, but likely requires other domains to recruit and/or regulate WASp/WAVE proteins and Rho family GTPases to further elaborate these actin-based scaffolds into dynamic structures (Fig. 5). Thus, Arg could simultaneously act as both a building block of an actin bundle scaffold and a regulator of the addition and dynamics of actin filaments arising from this scaffold. Future studies using fluorescence, video and electron microscopy should reveal the cytoskeletal ultrastructure and distribution of regulatory proteins in the Arg-containing protrusive structures.

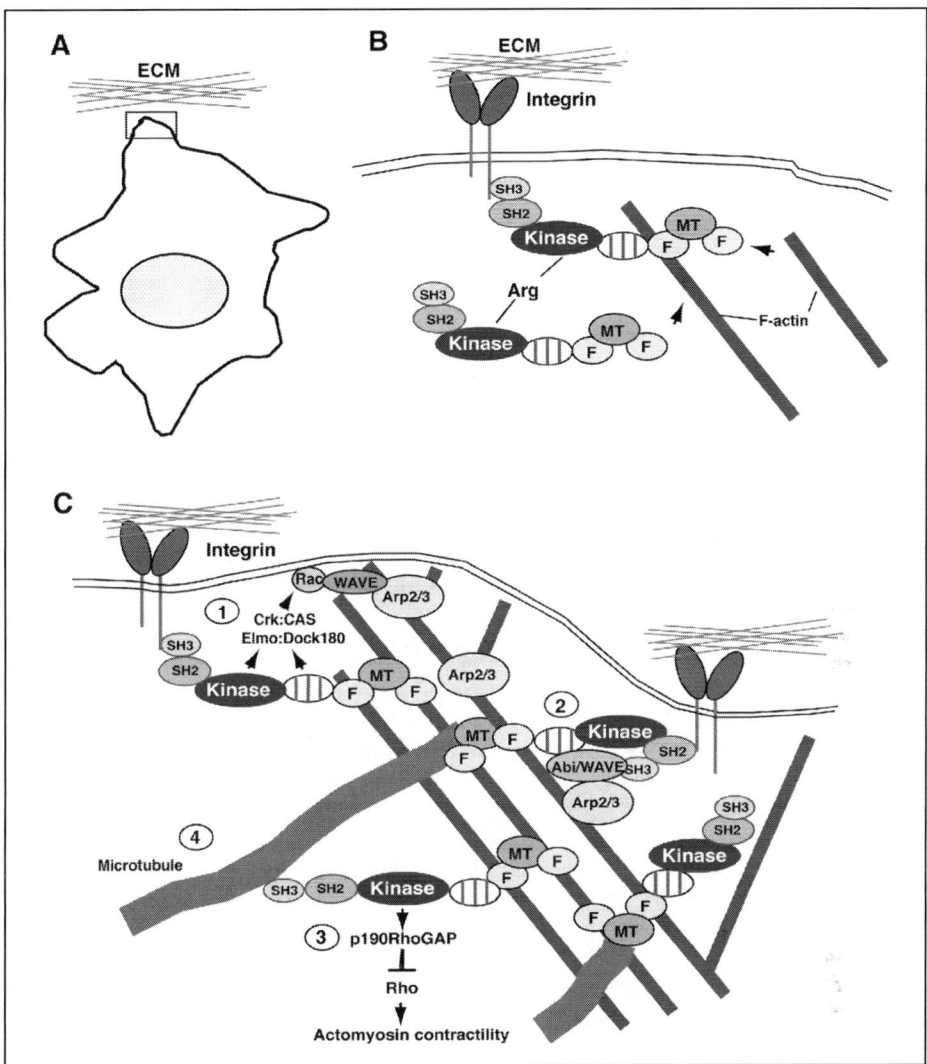

Figure 5. Arg promotes formation and assembly of an F-actin scaffold at protrusive sites. Integrin-mediated attachment leads to the formation of protrusive Arg- and F-actin-rich protrusive structures at the fibroblast periphery. The following model is proposed for the assembly and function of these structures. The depiction of Arg and its domains is identical to that in Figure 1. A) The fibroblast edge comes into contact with extracellular matrix (ECM). The resulting events at this leading edge (boxed area) are shown in B and C. B) Integrin receptors recruit Arg, which uses its F-actin-binding domains (F) to bind to F-actin.[16] Arg binding to F-actin stimulates the cooperative binding of additional Arg molecules to F-actin.[16] C) Arg organizes F-actin into a scaffold that serves as a platform for Arg to direct cytoskeletal rearrangements via several potential mechanisms. 1: Arg promotes assembly of a Crk:CAS:Elmo:Dock180 complex that activates Rac.[64] Activated Rac binds to the WAVE-containing complexes and promotes Arp2/3-dependent F-actin nucleation.[84,87] 2: Arg stimulates formation of Abi-1/WAVE2 complexes that activate Arp2/3-dependent F-actin nucleation.[89] 3: Arg phosphorylates p190RhoGAP, leading to inhibition of Rho and decreased actomyosin contractility. 4: Microtubules invade the scaffold and bind to the microtubule-binding domain of Arg.[17] Interactions with microtubules help localize Arg to the periphery. In addition, microtubule targeting to the scaffold may deliver proteins that regulate Arg-dependent processes.

Abl May Be Subject to Feedback Inhibition by F-Actin

Abl kinase activity is inhibited in fibroblasts that are held in suspension, and is induced upon attachment of cells to fibronectin.[35,105,106] An increased amount of F-actin coimmunoprecipitates with Abl from suspended cells relative to adherent cells. Treatment of cells with Latrunculin A, which reduces cellular F-actin concentration by blocking actin polymerization, leads to a potent induction of Abl kinase activity, even in suspended cells. Together, these observations strongly suggest that Abl kinase activity is influenced by differences in F-actin abundance or conformation in suspended versus adherent cells.

It has been proposed that F-actin inhibits Abl kinase activity by stabilizing an inhibited conformation of the enzyme.[107] Purified F-actin inhibits the ability of Abl to phosphorylate a peptide derived from C-terminal domain (CTD) of RNA polymerase II in vitro.[106] Both the F-actin-binding and SH2 domains of Abl are required for effect. However, F-actin does not inhibit the ability of Abl to phosphorylate its substrate Crk in vitro.[108] One possible explanation for this selective inhibitory effect is that F-actin competes with the CTD for binding to the Abl SH2 or F-actin-binding domains. The CTD binds with low micromolar affinity to an Abl fragment containing the F-actin binding domain.[109] Abl mutants lacking the SH2 or F-actin-binding domains do not bind to F-actin in vitro and therefore would not be inhibited by F-actin. Abl phosphorylation of Crk is not inhibited because Crk does not interact with the SH2 or F-actin binding domains.

An alternative scenario is that Abl exists in an inactive conformation in unattached cells and integrin-mediated adhesion promotes activation of Abl kinase activity. If this is the case, how does Latrunculin A treatment lead to activation of Abl kinase activity in the absence of adhesion? Latrunculin A treatment of suspended fibroblasts might activate adhesion receptors, mobilize calcium, or activate Rho family GTPases as it has been shown to do in other cell types.[110,111] One or more of these pathways could activate Abl kinase activity.

Abl Family Kinases Mediate Interactions between the F-Actin and Microtubule Networks

Polarized cell migration requires dynamic interactions between the actin and microtubule cytoskeletons.[112,113] MTs assume a polarized orientation in migrating cells, with their plus ends pointed toward the leading edge.[112,113] Microtubule extension into the periphery is required for lamellipodial protrusion at the leading edge.[114] Recent studies suggest that Abl family kinases might mediate interactions between F-actin and MTs at the leading edge.

Arg has a microtubule-binding domain located just between its two F-actin-binding domains.[17] Arg can bind MTs with high affinity and crosslink F-actin bundles to MTs in vitro.[17] These data suggest that Arg might mediate interactions between F-actin and MTs in cells (Fig. 5). In adhering fibroblasts, Arg localizes to the cell periphery and promotes the formation of F-actin-rich lamellipodial protrusions. MTs concentrate and insert into these structures.[17] Elimination of MTs with nocodazole disrupts Arg localization to the periphery, suggesting that Arg:MT interactions help concentrate Arg at protrusive sites. An Arg mutant lacking part of the MT-binding domain does not promote lamellipodial protrusions, even though it can still concentrate at peripheral sites. These data suggest that Arg:MT interactions are essential for Arg to promote lamellipodial protrusion. Arg might serve as part of a peripheral target for MTs.

The microtubule plus end binding protein CLASP binds to a subset of microtubules that interact with the actin-rich periphery in Xenopus growth cones.[115] Interestingly, its fly ortholog Orbit/MAST was identified as a regulator of midline axon guidance in flies. Importantly, genetic studies indicate that Orbit/MAST function is required for axon guidance defects resulting from Abl overexpression in the intersegmental nerve. These data indicate that a critical mediator or MT:F-actin interactions in the growth cone periphery acts downstream of Abl. Future studies should indicate how Abl modulates Orbit/MAST activity during growth cone guidance.

Conclusion

Abundant evidence implicates Abl family kinases as important links between cell surface receptors and the cellular machinery that regulates cytoskeletal rearrangements. In addition to the signaling partners of Abl family kinases discussed here, future studies are likely to identify additional pathways by which these kinases promote changes in cytoskeletal structure.

One big remaining challenge is to understand how Abl family kinases interface with their various effectors to produce cytoskeletal changes. Having identified many components of Abl kinase signaling pathways, we need to rigorously characterize their biochemical properties and examine how these properties are affected by interactions with Abl family kinases. We also need to pinpoint these activities in the cell and measure how they regulate F-actin and MT rearrangements. These efforts will benefit from ongoing advances in microscopy techniques, including both FRET-based probes to monitor biochemical activities and protein interactions and fluorescent speckle microscopy to examine actin and MT dynamics in live cells. The ultimate goal should be to understand Abl-controlled cytoskeletal structures well enough to model them in silico and make testable predictions about their properties.

In addition to activating mitogenic and anti-apoptotic pathways, oncogenic forms of Abl family kinases lead to derangement of cytoskeletal and adhesion pathways. A second challenge is to understand how Bcr-Abl-induced cytoskeletal changes contribute to disease phenotypes. These processes can be examined using mouse model systems of Bcr-Abl-positive leukemias (see chapter by Ren). Deletion of the F-actin binding domain reduces the oncogenic properties of the p190 form of Bcr-Abl. This observation suggests that disruption of cytoskeletal control pathways might be a useful strategy to treat Bcr-Abl-positive leukemias. The goal here will be to test whether these cytoskeletal pathways can be targeted for therapeutic purposes. Although Abl kinase inhibitors show tremendous promise for leukemia treatment, this treatment will probably be most useful in combination with other drugs or therapies.

Acknowledgements

I thank Ed Egelman, Vitoly Galkin, Mark Mooseker, Mark Peifer, David Van Vactor, for many helpful discussions and Stefanie Lapetina, Matt Miller, and Justin Peacock for comments on this Chapter. Work in my laboratory is supported by grants from the PHS (NS39475), the Kavli Institute for Neuroscience at Yale, and a Scholar Award from the Leukemia and Lymphoma Society of America.

References

1. Wang JY, Ledley F, Goff S et al. The mouse c-abl locus: Molecular cloning and characterization. Cell 1984; 36(2):349-56.
2. Kruh GD Perego R, Miki T et al. The complete coding sequence of arg defines the Abelson subfamily of cytoplasmic tyrosine kinases. Proc Natl Acad Sci USA 1990; 87(15):5802-6.
3. Henkemeyer MJ Bennett RL, Gertler FB et al. DNA sequence, structure, and tyrosine kinase activity of the Drosophila melanogaster Abelson proto-oncogene homolog. Mol Cell Biol 1988; 8(2):843-53.
4. Deng X, Hofmann ER, Villanueva A, Hobert O et al. Caenorhabditis elegans ABL-1 antagonizes p53-mediated germline apoptosis after ionizing irradiation. Nat Genet 2004; 36(8):906-12.
5. Barila D, Superti-Furga G. An intramolecular SH3-domain interaction regulates c-Abl activity. Nat Genet 1998; 18(3):280-2.
6. Nagar B, Hantschel O, Young MA et al. Structural basis for the autoinhibition of c-Abl tyrosine kinase. Cell 2003; 112(6):859-71.
7. Mayer BJ, Baltimore D. Mutagenic analysis of the roles of SH2 and SH3 domains in regulation of the Abl tyrosine kinase. Mol Cell Biol 1994; 14(5):2883-94.
8. Mayer BJ, Hirai H, Sakai R. Evidence that SH2 domains promote processive phosphorylation by protein-tyrosine kinases. Curr Biol 1995; 5(3):296-305.
9. Ren R, Ye ZS, Baltimore D. Abl protein-tyrosine kinase selects the Crk adapter as a substrate using SH3-binding sites. Genes Dev 1994; 8(7):783-95.
10. Pendergast AM. The Abl family kinases: Mechanisms of regulation and signaling. Adv Cancer Res 2002; 85:51-100.

11. Shi Y, Alin K, Goff SP. Abl-interactor-1, a novel SH3 protein binding to the carboxy-terminal portion of the Abl protein, suppresses v-abl transforming activity. Genes Dev 1995; 9(21):2583-97.

12. Dai Z, Pendergast AM. Abi-2, a novel SH3-containing protein interacts with the c-Abl tyrosine kinase and modulates c-Abl transforming activity. Genes Dev 1995; 9(21):2569-82.

13. Wang B, Mysliwiec T, Krainc D et al. Identification of ArgBP1, an Arg protein tyrosine kinase binding protein that is the human homologue of a CNS-specific Xenopus gene. Oncogene 1996; 12(9):1921-9.

14. McWhirter JR, Wang JY. An actin-binding function contributes to transformation by the Bcr-Abl oncoprotein of Philadelphia chromosome-positive human leukemias. EMBO J 1993; 12(4):1533-46.

15. Van Etten RA, Jackson PK, Baltimore D et al. The COOH terminus of the c-Abl tyrosine kinase contains distinct F- and G-actin binding domains with bundling activity. J Cell Biol 1994; 124(3):325-40.

16. Wang Y, Miller AL, Mooseker MS et al. The Abl-related gene (Arg) nonreceptor tyrosine kinase uses two F-actin-binding domains to bundle F-actin. Proc Natl Acad Sci USA 2001; 98(26):14865-70.

17. Miller AL, Wang Y, Mooseker MS et al. The Abl-related gene (Arg) requires its F-actin-microtubule cross-linking activity to regulate lamellipodial dynamics during fibroblast adhesion. J Cell Biol 2004; 165(3):407-19.

18. Miao YJ, Wang JY. Binding of A/T-rich DNA by three high mobility group-like domains in c-Abl tyrosine kinase. J Biol Chem 1996; 271(37):22823-30.

19. Wen ST, Jackson PK, Van Etten RA. The cytostatic function of c-Abl is controlled by multiple nuclear localization signals and requires the p53 and Rb tumor suppressor gene products. EMBO J 1996; 15(7):1583-95.

20. Taagepera S, McDonald D, Loeb JE et al. Nuclear-cytoplasmic shuttling of C-ABL tyrosine kinase. Proc Natl Acad Sci USA 1998; 95(13):7457-62.

21. Melo JV. BCR-ABL gene variants. Baillieres Clin Haematol 1997; 10(2):203-22.

22. McWhirter JR, Galasso DL, Wang JY. A coiled-coil oligomerization domain of Bcr is essential for the transforming function of Bcr-Abl oncoproteins. Mol Cell Biol 1993; 13(12):7587-95.

23. Zhao X, Ghaffari S, Lodish H et al. Structure of the Bcr-Abl oncoprotein oligomerization domain. Nat Struct Biol 2002; 9(2):117-20.

24. McWhirter JR, Wang JY. Activation of tyrosinase kinase and microfilament-binding functions of c-abl by bcr sequences in bcr/abl fusion proteins. Mol Cell Biol 1991; 11(3):1553-65.

25. Muller AJ, Young JC, Pendergast AM et al. BCR first exon sequences specifically activate the BCR/ABL tyrosine kinase oncogene of Philadelphia chromosome-positive human leukemias. Mol Cell Biol 1991; 11(4):1785-92.

26. Smith KM, Yacobi R, Van Etten RA. Autoinhibition of Bcr-Abl through its SH3 domain. Mol Cell 2003; 12(1):27-37.

27. Raitano AB, Whang YE, Sawyers CL. Signal transduction by wild-type and leukemogenic Abl proteins. Biochim Biophys Acta 1997; 1333(3):F201-16.

28. Chopra R, Pu QQ, Elefanty AG. Biology of BCR-ABL. Blood Rev 1999; 13(4):211-29.

29. Ren R. Mechanisms of BCR-ABL in the pathogenesis of chronic myelogenous leukaemia. Nat Rev Cancer 2005; 5(3):172-83.

30. Salgia R, Li JL, Ewaniuk DS et al. BCR/ABL induces multiple abnormalities of cytoskeletal function. J Clin Invest 1997; 100(1):46-57.

31. Weisberg E, Sattler M, Ewaniuk DS et al. Role of focal adhesion proteins in signal transduction and oncogenesis. Crit Rev Oncog 1997; 8(4):343-58.

32. Wennstrom S, Siegbahn A, Yokote K et al. Membrane ruffling and chemotaxis transduced by the PDGF beta-receptor require the binding site for phosphatidylinositol 3' kinase. Oncogene 1994; 9(2):651-60.

33. Plattner R, Kadlec L, DeMali KA et al. c-Abl is activated by growth factors and Src family kinases and has a role in the cellular response to PDGF. Genes Dev 1999; 13(18):2400-11.

34. Sini P, Cannas A, Koleske AJ et al. Abl-dependent tyrosine phosphorylation of Sos-1 mediates growth-factor-induced Rac activation. Nat Cell Biol 2004; 6(3):268-74.

35. Woodring PJ, Litwack ED, O'Leary DD et al. Modulation of the F-actin cytoskeleton by c-Abl tyrosine kinase in cell spreading and neurite extension. J Cell Biol 2002; 156(5):879-92.

36. Woodring PJ, Meisenhelder J, Johnson SA et al. c-Abl phosphorylates Dok1 to promote filopodia during cell spreading. J Cell Biol 2004; 165(4):493-503.

37. Grevengoed EE, Loureiro JJ, Jesse TL et al. Abelson kinase regulates epithelial morphogenesis in Drosophila. J Cell Biol 2001; 155(7):1185-98.

38. Burnside B. Microtubules and microfilaments in newt neurulation. Dev Biol 1971; 26:416-441.

39. Burnside B. Microtubules and microfilaments in amphibian neurulation. Amer Zool 1973; 13:989-1006.
40. Karfunkel P. The activity of microtubules and microfilaments in neurulation in the chick. J Exp Zool 1971; 181:289-302.
41. Baker PC, Schroeder TE. Cytoplasmic filaments and morphogenetic movement in the amphibian neural tube. Dev Biol 1967; 15:432-450.
41a. Koleske AJ, Gifford AM, Scott ML et al. Essential roles for the Abl and Arg tyrosine kinases in neurulation. Neuron 1998; 21:1259-1272.
42. Grevengoed EE, Fox DT, Gates J et al. Balancing different types of actin polymerization at distinct sites: Roles for Abelson kinase and Enabled. J Cell Biol 2003; 163(6):1267-79.
43. Bennett RL, Hoffmann FM. Increased levels of the Drosophila Abelson tyrosine kinase in nerves and muscles: Subcellular localization and mutant phenotypes imply a role in cell-cell interactions. Development 1992; 116(4):953-66.
44. Adam T, Arpin M, Prevost MC et al. Cytoskeletal rearrangements and the functional role of T-plastin during entry of Shigella flexneri into HeLa cells. J Cell Biol 1995; 129(2):367-81.
45. Burton EA, Plattner R, Pendergast AM. Abl tyrosine kinases are required for infection by Shigella flexneri. EMBO J 2003; 22(20):5471-9.
46. Hollinshead M, Rodger G, Van Eijl H et al. Vaccinia virus utilizes microtubules for movement to the cell surface. J Cell Biol 2001; 154(2):389-402.
47. Rietdorf J, Ploubidou A, Reckmann I et al. Kinesin-dependent movement on microtubules precedes actin-based motility of vaccinia virus. Nat Cell Biol 2001; 3(11):992-1000.
48. Newsome TP, Scaplehorn N, Way M. SRC mediates a switch from microtubule- to actin-based motility of vaccinia virus. Science 2004; 306(5693):124-9.
49. Reeves PM, Bommarius B, Lebeis S et al. Disabling poxvirus pathogenesis by inhibition of Abl-family tyrosine kinases. Nat Med 2005.
50. Kaibuchi K, Kuroda S, Amano M. Regulation of the cytoskeleton and cell adhesion by the Rho family GTPases in mammalian cells. Annu Rev Biochem 1999; 68:459-86.
51. Ridley AJ. Rho family proteins: Coordinating cell responses. Trends Cell Biol 2001; 11(12):471-7.
52. Burridge K, Wennerberg K. Rho and Rac take center stage. Cell 2004; 116(2):167-79.
53. Ridley AJ, Hall A. The small GTP-binding protein rho regulates the assembly of focal adhesions and actin stress fibers in response to growth factors. Cell 1992; 70(3):389-99.
54. Ridley AJ, Paterson HF, Johnston CL et al. The small GTP-binding protein rac regulates growth factor-induced membrane ruffling. Cell 1992; 70(3):401-10.
55. Kozma R, Ahmed S, Best A et al. The Ras-related protein Cdc42Hs and bradykinin promote formation of peripheral actin microspikes and filopodia in Swiss 3T3 fibroblasts. Mol Cell Biol 1995; 15(4):1942-52.
56. Nobes CD, Hall A. Rho, rac, and cdc42 GTPases regulate the assembly of multimolecular focal complexes associated with actin stress fibers, lamellipodia, and filopodia. Cell 1995; 81(1):53-62.
57. Pollard TD, Borisy GG. Cellular motility driven by assembly and disassembly of actin filaments. Cell 2003; 112(4):453-65.
58. Nobes CD, Hawkins P, Stephens L et al. Activation of the small GTP-binding proteins rho and rac by growth factor receptors. J Cell Sci 1995; 108(Pt 1):225-33.
59. Plattner R, Koleske AJ, Kazlauskas A et al. Bidirectional signaling links the Abelson kinases to the platelet-derived growth factor receptor. Mol Cell Biol 2004; 24(6):2573-83.
60. Skorski T, Wlodarski P, Daheron L et al. BCR/ABL-mediated leukemogenesis requires the activity of the small GTP-binding protein Rac. Proc Natl Acad Sci USA 1998; 95(20):11858-62.
61. Renshaw MW, Lea-Chou E, Wang JY. Rac is required for v-Abl tyrosine kinase to activate mitogenesis. Curr Biol 1996; 6(1):76-83.
62. Scita G, Nordstrom J, Carbone R et al. EPS8 and E3B1 transduce signals from Ras to Rac. Nature 1999; 401(6750):290-3.
63. DeMali KA, Wennerberg K, Burridge K. Integrin signaling to the actin cytoskeleton. Curr Opin Cell Biol 2003; 15(5):572-82.
64. Chodniewicz D, Klemke RL. Regulation of integrin-mediated cellular responses through assembly of a CAS/Crk scaffold. Biochim Biophys Acta 2004; 1692(2-3):63-76.
65. Klemke RL, Leng J, Molander R et al. CAS/Crk coupling serves as a "molecular switch" for induction of cell migration. J Cell Biol 1998; 140(4):961-72.
66. Matsuda M, Ota S, Tanimura R et al. Interaction between the amino-terminal SH3 domain of CRK and its natural target proteins. J Biol Chem 1996; 271(24):14468-72.
67. Feller SM, Knudsen B, Hanafusa H. c-Abl kinase regulates the protein binding activity of c-Crk. EMBO J 1994; 13(10):2341-51.

68. Wang B, Mysliwiec T, Feller SM et al. Proline-rich sequences mediate the interaction of the Arg protein tyrosine kinase with Crk. Oncogene 1996; 13(7):1379-85.
69. Kain KH, Klemke RL. Inhibition of cell migration by Abl family tyrosine kinases through uncoupling of Crk-CAS complexes. J Biol Chem 2001; 276(19):16185-92.
70. Escalante M, Courtney J, Chin WG et al. Phosphorylation of c-Crk II on the negative regulatory Tyr222 mediates nerve growth factor-induced cell spreading and morphogenesis. J Biol Chem 2000; 275(32):24787-97.
71. Hernandez SE, Settleman J, Koleske AJ. Adhesion-dependent regulation of p190RhoGAP in the developing brain by the Abl-related gene tyrosine kinase. Curr Biol 2004; 14(8):691-6.
72. Jones SB, Lu HY, Lu Q. Abl tyrosine kinase promotes dendrogenesis by inducing actin cytoskeletal rearrangements in cooperation with Rho family small GTPases in hippocampal neurons. J Neurosci 2004; 24(39):8510-21.
73. Moresco EM, Donaldson S, Williamson A et al. Integrin-mediated dendrite branch maintenance requires Abelson (Abl) family kinases. J Neurosci 2005; 25(26):6105-18.
74. Hu KQ, Settleman J. Tandem SH2 binding sites mediate the RasGAP-RhoGAP interaction: A conformational mechanism for SH3 domain regulation. EMBO J 1997; 16(3):473-83.
75. Roof RW, Haskell MD, Dukes BD et al. Phosphotyrosine (p-Tyr)-dependent and -independent mechanisms of p190 RhoGAP-p120 RasGAP interaction: Tyr 1105 of p190, a substrate for c-Src, is the sole p-Tyr mediator of complex formation. Mol Cell Biol 1998; 18(12):7052-63.
76. Chang JH, Gill S, Settleman J et al. c-Src regulates the simultaneous rearrangement of actin cytoskeleton, p190RhoGAP, and p120RasGAP following epidermal growth factor stimulation. J Cell Biol 1995; 130(2):355-68.
77. Haskell MD, Nickles AL, Agati JM et al. Phosphorylation of p190 on Tyr1105 by c-Src is necessary but not sufficient for EGF-induced actin disassembly in C3H10T1/2 fibroblasts. J Cell Sci 2001; 114(Pt 9):1699-708.
78. Brouns MR, Matheson SF, Settleman J. p190 RhoGAP is the principal Src substrate in brain and regulates axon outgrowth, guidance and fasciculation. Nat Cell Biol 2001; 3(4):361-7.
79. Arthur WT, Petch LA, Burridge K. Integrin engagement suppresses RhoA activity via a c-Src-dependent mechanism. Curr Biol 2000; 10(12):719-22.
80. Dorey K, Engen JR, Kretzschmar J et al. Phosphorylation and structurebased functional studies reveal a positive and a negative role for the activation loop of the c-Abl tyrosine kinase. Oncogene 2001; 20(56):8075-84.
81. Tanis KQ, Veach D, Duewel HS et al. Two distinct phosphorylation pathways have additive effects on Abl family kinase activation. Mol Cell Biol 2003; 23(11):3884-96.
82. Schindler T, Bornmann W, Pellicena P et al. Structural mechanism for STI-571 inhibition of abelson tyrosine kinase. Science 2000; 289(5486):1938-42.
83. Stradal T, Courtney KD, Rottner K et al. The Abl interactor proteins localize to sites of actin polymerization at the tips of lamellipodia and filopodia. Curr Biol 2001; 11(11):891-5.
84. Eden S, Rohatgi R, Podtelejnikov AV et al. Mechanism of regulation of WAVE1-induced actin nucleation by Rac1 and Nck. Nature 2002; 418(6899):790-3.
85. Echarri A, Lai MJ, Robinson MR et al. Abl interactor 1 (Abi-1) wave-binding and SNARE domains regulate its nucleocytoplasmic shuttling, lamellipodium localization, and wave-1 levels. Mol Cell Biol 2004; 24(11):4979-93.
86. Kunda P, Craig G, Dominguez V et al. Abi, Sra1, and Kette control the stability and localization of SCAR/WAVE to regulate the formation of actin-based protrusions. Curr Biol 2003; 13(21):1867-75.
87. Innocenti M, Zucconi A, Disanza A et al. Abi1 is essential for the formation and activation of a WAVE2 signalling complex. Nat Cell Biol 2004; 6(4):319-27.
88. Steffen A, Rottner K, Ehinger J et al. Sra-1 and Nap1 link Rac to actin assembly driving lamellipodia formation. EMBO J 2004; 23(4):749-59.
89. Leng Y, Zhang J, Badour K et al. Abelson-interactor-1 promotes WAVE2 membrane translocation and Abelson-mediated tyrosine phosphorylation required for WAVE2 activation. Proc Natl Acad Sci USA 2005; 102(4):1098-103.
90. Cestra G, Toomre D, Chang S et al. The Abl/Arg substrate ArgBP2/nArgBP2 coordinates the function of multiple regulatory mechanisms converging on the actin cytoskeleton. Proc Natl Acad Sci USA 2005; 102(5):1731-6.
91. Yamanashi Y, Baltimore D. Identification of the Abl- and rasGAP-associated 62 kDa protein as a docking protein, Dok. Cell 1997; 88(2):205-11.
92. Carpino N, Wisniewski D, Strife A et al. p62(dok): A constitutively tyrosine-phosphorylated, GAP-associated protein in chronic myelogenous leukemia progenitor cells. Cell 1997; 88(2):197-204.

93. Master Z et al. Dok-R binds c-Abl and regulates Abl kinase activity and mediates cytoskeletal reorganization. J Biol Chem 2003; 278(32):30170-9.
94. Rivero-Lezcano OM et al. Wiskott-Aldrich syndrome protein physically associates with Nck through Src homology 3 domains. Mol Cell Biol 1995; 15(10):5725-31.
95. Rohatgi R et al. Nck and phosphatidylinositol 4,5-bisphosphate synergistically activate actin polymerization through the N-WASP-Arp2/3 pathway. J Biol Chem 2001; 276(28):26448-52.
96. Zhao ZS, Manser E, Lim L. Interaction between PAK and nck: A template for Nck targets and role of PAK autophosphorylation. Mol Cell Biol 2000; 20(11):3906-17.
97. Svitkina TM, Bulanova EA, Chaga OY et al. Mechanism of filopodia initiation by reorganization of a dendritic network. J Cell Biol 2003; 160(3):409-21.
98. Krause M, Dent EW, Bear JE et al. Ena/VASP proteins: Regulators of the actin cytoskeleton and cell migration. Annu Rev Cell Dev Biol 2003; 19:541-64.
99. Gertler FB, Doctor JS, Hoffmann FM. Genetic suppression of mutations in the Drosophila abl proto-oncogene homolog. Science 1990; 248(4957):857-60.
100. Comer AR, Ahern-Djamali SM, Juang JL et al. Phosphorylation of Enabled by the Drosophila Abelson tyrosine kinase regulates the in vivo function and protein-protein interactions of Enabled. Mol Cell Biol 1998; 18(1):152-60.
101. Tani K, Sato S, Sukezane T et al. Abl interactor 1 promotes tyrosine 296 phosphorylation of mammalian enabled (Mena) by c-Abl kinase. J Biol Chem 2003; 278(24):21685-92.
102. Juang JL, Hoffmann FM. Drosophila abelson interacting protein (dAbi) is a positive regulator of abelson tyrosine kinase activity. Oncogene 1999; 18(37):5138-47.
103. Galkin VE, Orlova A, Koleske AJ et al. The Arg nonreceptor tyrosine kinase modifies F-actin structure. J Mol Biol 2005; 346(2):565-75.
104. Koleske AJ, Gifford AM, Scott ML et al. Essential roles for the Abl and Arg tyrosine kinases in neurulation. Neuron 1998; 21(6):1259-72.
105. Lewis JM, Baskaran R, Taagepera S et al. Integrin regulation of c-Abl tyrosine kinase activity and cytoplasmic-nuclear transport. Proc Natl Acad Sci USA 1996; 93(26):15174-9.
106. Woodring PJ, Hunter T, Wang JY. Inhibition of c-Abl tyrosine kinase activity by filamentous actin. J Biol Chem 2001; 276(29):27104-10.
107. Wang JY. Controlling Abl: Auto-inhibition and coinhibition? Nat Cell Biol 2004; 6(1):3-7.
108. Woodring PJ, Hunter T, Wang JY. Mitotic phosphorylation rescues Abl from F-actin-mediated inhibition. J Biol Chem 2005; 280(11):10318-25.
109. Baskaran R, Chiang GG, Wang JY. Identification of a binding site in c-Abl tyrosine kinase for the C-terminal repeated domain of RNA polymerase II. Mol Cell Biol 1996; 16(7):3361-9.
110. Lim D, Lange K, Santella L. Activation of oocytes by latrunculin A. Faseb J 2002; 16(9):1050-6.
111. Zimmerman AW, Nelissen JM, van Emst-de Vries SE et al. Cytoskeletal restraints regulate homotypic ALCAM-mediated adhesion through PKCalpha independently of Rho-like GTPases. J Cell Sci 2004; 117(Pt 13):2841-52.
112. Vasiliev JM. Polarization of pseudopodial activities: Cytoskeletal mechanisms. J Cell Sci 1991; 98(Pt 1):1-4.
113. Bershadsky AD, Vaisberg EA, Vasiliev JM. Pseudopodial activity at the active edge of migrating fibroblast is decreased after drug-induced microtubule depolymerization. Cell Motil Cytoskeleton 1991; 19(3):152-8.
114. Waterman-Storer CM, Worthylake RA, Liu BP et al. Microtubule growth activates Rac1 to promote lamellipodial protrusion in fibroblasts. Nat Cell Biol 1999; 1(1):45-50.
115. Lee H, Engel U, Rusch J et al. The microtubule plus end tracking protein Orbit/MAST/CLASP acts downstream of the tyrosine kinase Abl in mediating axon guidance. Neuron 2004; 42(6):913-26.
116. Hernandez SE, Krishnaswami M, Miller AL et al. How do Abl family kinases regulate cell shape and movement? Trends Cell Biol 2004; 14(1):36-44.
117. Howe AK, Hogan BP, and Julicano RL. Regulation of vasoditator-stimulated phosphoprotein phosphorylation and interaction with Abl by protein kinase A and cell adhesion. J Biol Chem 2002; 277(41) 33121-38126.

CHAPTER 6

Regulation of Cell Motility by Abl Family Kinases

Shahin Emami and Richard L. Klemke*

Abstract

Cell migration is a highly dynamic process that involves regulation of actin-mediated protrusion of a leading lamellipodia and its adhesion to the extracellular matrix, followed by translocation of the cell body and tail retraction at the rear. The migration machinery is regulated in a highly temporal and spatial manner through sophisticated sensing mechanisms that interpret external gradients of chemokines and adhesive proteins present in the extracellular environment. These directional cues are transmitted to the interior of the cell where they couple to the cytoskeletal network. In the following section, we highlight the role of the Abl family of nonreceptor tyrosine kinases, which transmit signals from growth factor and adhesion receptors to the actin and microtubule cytoskeleton of motile cells. These recent findings suggest that Abl kinases may contribute to cell migration processes, including development, wound healing, and immune function, as well as pathological conditions associated with cancer metastasis and inflammation.

Introduction

Cell migration is essential for normal development, immune function, and wound healing. However, when deregulated, cell migration contributes to autoimmune disease, inflammation, tumor-induced angiogenesis, and cancer cell metastasis. Regulation of cell movement is a complex process that involves cell adhesion to extracellular matrix proteins and gradients of growth factors present in the extracellular environment, which serve to guide cell movement through complex tissues. Cell migration is an integrated process that involves sensing directional cues from the extracellular environment, regulation of actin-mediated protrusion of a dominant leading lamellipodium at the front, tail retraction at the rear, and turnover of cell-matrix adhesions to the underlying substratum.[1,2] All these events are coordinately controlled by complex signaling networks that operate in a temporal and spatial manner within the cell.[2]

Much of our knowledge of cell migration has been obtained from cultured cell lines and 2-dimensional in vitro assays, which do not fully recapitulate true 3-dimensional cell translocation that occurs in living organisms. Disruption of normal tissue architecture and establishment of cells in culture leads to reactivation of dormant migration mechanisms that would normally be suppressed since most cells in the adult body do not migrate except under special circumstances associated with immune function and wound healing.[1,2] However, significant progress has been made in understanding the fundamental process of cell movement and a detailed understanding of the molecular events that transform a stationary cell into fully

*Corresponding author: Richard L. Klemke—Department of Immunology, The Scripps Research Institute, 10550 North Torrey Pines Road, SP231, La Jolla, California, 92037, U.S.A. Email: klemke@scripps.edu.

Abl Family Kinases in Development and Disease, edited by Anthony Koleske.
©2006 Landes Bioscience and Springer Science+Business Media.

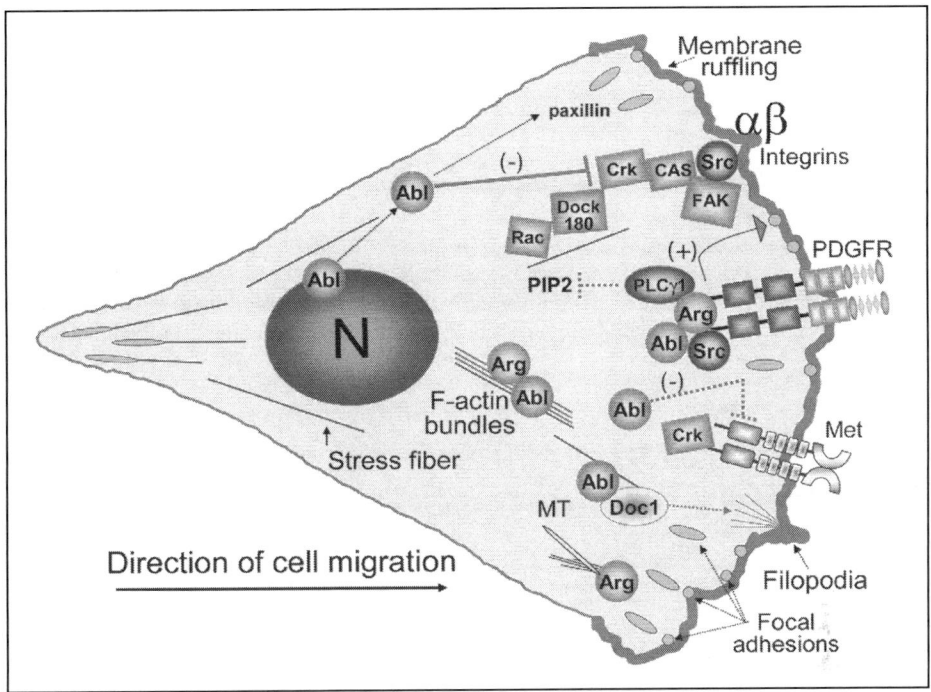

Figure 1. Schematic of a migrating cell and proposed regulation by Abl kinases. Growth factors and adhesion proteins in the extracellular environment regulate cell migration through stimulation of integrins and growth factor receptors. Cell migration is a dynamic process that involves protrusion of a leading lamellipodium with actin-rich membrane ruffles, membrane attachment to the ECM and formation of focal adhesions at the front, and simultaneous detachment of adhesive contacts at the rear of the cell. Current evidence indicates that Abl and Arg nonreceptor tyrosine kinases control cell motility through regulation of signal transduction processes downstream of integrin and growth factor receptors, including c-Met and PDGF receptors.[22,28,40] Abl and Arg also bind to F-actin structures as well as microtubules (Arg), suggesting additional mechanisms for regulation of cell migration.[9,42] Abelson tyrosine kinase (Abl), Abl-related gene (Arg), α and β integrin receptors (αβ), Crk-associated substrate (CAS), CT10 regulator of kinase (Crk), focal adhesion kinase (FAK), hepatocyte growth factor (HGF), hepatocyte growth factor receptor (Met), microtubules (MT), nucleus (N), phospholipase C gamma one (PLC-γ1), platelet-derived growth factor receptor (PDGFR), phosphatidyl inositol 4,5-bisphosphate (PIP2).

functioning migratory cell is beginning to take shape. Progress is also being made in extending these findings into 3-dimensional systems in vitro and in vivo.[3,4] Here we highlight the importance of the Abelson family of non receptor tyrosine kinases (c-Abl and the c-Abl-related gene, Arg), which have recently emerged as key regulators of cell migration through their ability to relay signals from growth factor and integrin adhesion receptors to the actin and microtubule cytoskeleton that controls the migration machinery of the cell (Fig. 1).

Domain Structure and Developmental Function of Abl Family Kinases

The Abl family kinases are evolutionarily conserved non receptor tyrosine kinases that include the normal cellular c-Abl (Abl) and Arg, and the oncogenic forms v-Abl, Bcr-Abl, Tel-Arg.[5-10] While a large body of literature describes the importance of the oncogenic forms in human health and disease, we will focus on the emerging role of endogenous Abl and Arg in regulation of cell movement in mammalian cells. Abl and Arg are ubiquitously expressed proteins that share high homology, comprised of SH2, SH3, actin binding, polyproline, and tyrosine kinase domains.[5,7-9] Not surprisingly they share a significant level of functional redundancy and utilize similar effector proteins.[7] However, a notable difference in these proteins is the ability of c-Abl to shuttle between the nucleus and cytoplasm due to nuclear localization signals (NLS), which are missing from the Arg gene product.[8,10] Also, while approximately 75% of *abl-/-* animals die postpartum, the surviving mice show thymic atrophy, lymphopenia, osteoporosis, improper eye development, and spermatogenesis.[11-14] In contrast, *arg-/-* mice are viable, but show behavioral defects. When both *abl* and *arg* genes are deleted, animals die early in development (E9-11) from neurological defects, indicating the necessity of both enzymes for proper embryonic development and that they can compensate for each other during normal development.[12] Since proper cell migration is necessary for normal development, it is likely that the early lethality observed in the *abl-/-arg-/-* cells is at least in part due to deregulated migration events, especially during neural tube formation, in which Abl/Arg play a key role.[12] This may also explain the high level of apoptosis seen in the Abl/Arg null embryos since defective migratory cells that incorrectly colonize developing tissues during pattern formation are readily eliminated by apoptosis.

Negative Regulation of Cell Migration by Abl Kinases through Uncoupling of CAS/Crk Complexes

Signal transduction by Abl kinases plays a key role during cell migration by virtue of their ability to integrate a diverse repertoire of signals that emanate from the extracellular environment, including extracellular matrix proteins (ECM) and growth factors.[9,15,16] Upon integrin ligation during cell adhesion to ECM proteins and/or exposure to cytokines, Abl and Arg are reported to be activated and localized to focal adhesions and actin-rich membrane ruffles.[17-19] These early reports were the first hint that Abl members may regulate the cytoskeleton and cell migration processes. Subsequently, Abl was shown to negatively regulate cell migration by disrupting the formation of a macromolecular signaling complex between the adaptor protein c-CrkII (Crk) and the docking protein p130CAS (Crk-associated substrate, CAS) and its effector proteins DOCK180 and the small GTPase Rac, which control the actin cytoskeleton and lamellipodium formation (refs. 20-22, Figs. 2, 3). Abl and Arg directly interact with the SH3 domain of Crk via their conserved proline-rich region (PXXP, Fig. 2) and promote tyrosine phosphorylation of Crk at Y221.[21-25] Y221 phosphorylation causes an intramolecular folding in which Crk's own SH2 domain binds to Y221 preventing interaction with phosphotyrosine residues present in the substrate domain of CAS as well as interactions with effector proteins that bind to the amino terminal SH3 domain of Crk. Thus, the binding of Abl to Crk is expected to inhibit cell migration by both phosphorylation of Y221 and displacement of effectors that bind to Crk's SH3 domain, including DOCK180.[20-22]

Consistent with this idea, expression of a kinase dead form of Abl or exposure of cells to the Abl kinase inhibitor STI571[21,22,26] enhanced CAS/Crk coupling and cell migration by decreasing phosphorylation of CrkY221.[22] Conversely, expression of a constitutively activated form of Abl inhibited cell migration and the assembly of CAS/Crk complexes, and this depended on a functional Abl PXXP domain and phosphorylation of CrkY221.[22,24,25] Furthermore, embryonic fibroblast cells (MEF) isolated from *abl-/-arg-/-* deficient mice showed a complete loss of basal CrkY221 phosphorylation that led to increased CAS/Crk assembly and cell migration. As expected, reconstitution of these cells with Abl reversed this phenotype.[22] Interestingly, *abl-/-arg+/+* MEF cells show increased CrkY221 phosphorylation, indicating that Arg can also promote phosphorylation of endogenous Crk. Collectively, these studies demonstrated that Abl and Arg kinases function as negative regulators of CAS/Crk coupling and cell migration. This also strongly implicates endogenous Abl/Arg kinases as the primary regulators of Crk phosphorylation in cells and points to an important "housekeeping" role to monitor Crk phosphorylation levels in cells. The highly conserved CAS/Crk/DOCK180/Rac signaling module has been shown to regulate growth factor and integrin mediated cell migration in various cell types and in cell migration processes in vivo and, when deregulated, may contribute to cancer progression.[21,27,28]

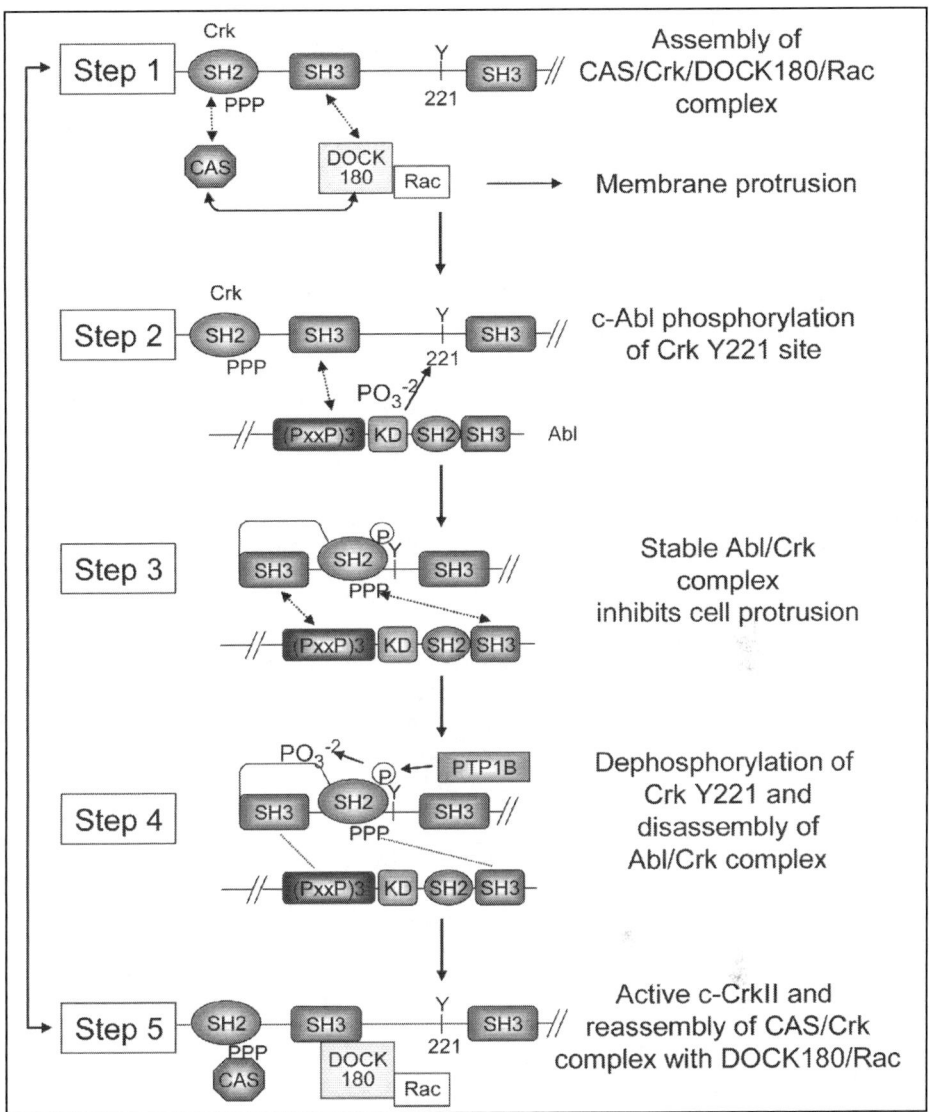

Figure 2. Proposed mechanism of Abl regulation of CAS/Crk coupling and membrane protrusion. Step 1) Growth factor and integrin receptor activation facilitates the assembly of a CAS/Crk/DOCK180/Rac signaling scaffold that promotes actin-mediated lamellipodium protrusion.[20,21] Step 2) Abl is also activated under these conditions and serves as a negative-feedback signal that modulates the level of CAS/Crk coupling in cells through phosphorylation of tyrosine 221 (Y221) in Crk.[22] Step 3) Phosphorylated Crk then undergoes intramolecular folding in which its own SH2 domain binds to phosphorylated Y221. This in turn exposes a proline rich region (PXXP) within the Crk-SH2 domain, which can interact with the SH3 domain of Abl forming a stable complex.[29] Step 4) The Abl/Crk complex is then translocated to the membrane and/or focal adhesions where it is dephosphorylated by protein tyrosine phosphatase-1B (PIP-1B).[21] Step 5) This facilitates the disassembly of Abl and Crk, enabling Crk to unfold and interact with the CAS protein scaffold. The cycle is then repeated as new protrusions and focal adhesions form in the leading lamellipodium during cell migration. Src homology 2 (SH2), Src homology 3 (SH3), polyproline sequence (PPP), kinase domain (KD).

Abl activation by integrins and growth factor receptors may serve as a negative feedback pathway to control the level of CAS/Crk coupling by reducing the pool of active unphosphorylated Crk available to interact with CAS following integrin activation. In migrating cells, this negative feedback signal would be expected to operate transiently downstream of integrin ligation events occurring at the leading front of the advancing lamellipodium in order to bring about steady-state levels of CAS/Crk complexes in this structure.[4,21] Also, as indicated above, the increased association of Abl with the SH3 domain of Crk may displace positive effectors like DOCK180 to further reinforce the inhibition of cell migration by Abl. Interestingly, Abl phosphorylation and molecular folding of Crk has been shown to unmask a polyproline sequence (PPP) motif in Crk that can further interact with the SH3 domain of Abl, (ref. 29, Fig. 2). The increased molecular stability of this protein complex may provide a mechanism to compete with positive effectors for binding to the SH3 domain of Crk, which could further potentiate the negative regulation of cell migration by Abl.[29,30]

The ability of Abl to both phosphorylate and tightly bind to Crk as well as its ability to translocate within the cell suggests that this enzyme could also serve as a shuttle protein to deliver phosphorylated Crk to the membrane and/or focal adhesions (Fig. 2).[17,18,31,32] In this case, integrin-induced signals would induce formation of Abl/Crk complexes and their translocation to the membrane and/or focal contacts. In this way, integrin-mediated activation of Abl would provide a sensitive mechanism to modulate the pool of active/inactive Crk available to bind to effectors in distinct compartments of the cell. Further studies will be necessary to determine whether Crk retains the ability to translocate to the membrane and/or focal contacts in *abl-/-arg-/-* cells and whether integrin-mediated Rac activation is compromised in cells without functional Abl family kinases. Also, it will be important to identify the upstream components responsible for integrin activation of Abl family kinases during cell spreading and migration. Src would be a good candidate for this role, as it is activated by integrins and can in turn activate Abl and regulate cell migration.[33,34]

Abl Regulation of HGF-Induced Cell Migration

Inhibition of Abl kinase activity by STI571 has also been shown to enhance hepatocyte growth factor (HGF)-induced motility in a panel of thyroid cancer cell lines.[28] In this study, treatment of cells with clinically relevant doses of STI571 increased HGF-induced cell migration in 10 of 13 independently derived cell lines and induced strong branching morphogenesis and cell invasion in Matrigel. Inhibition of basal Abl kinase activity with STI571 in the absence of HGF also increased cell migration in 9 of 13 cancer cell lines and induced a moderate increase in branching and cell invasion. Interestingly, STI571 had no effect on primary thyroid cell lines, suggesting Abl kinase activity specifically and negatively regulates cell migration and invasion in malignant thyroid cancer cells.

Although the mechanism of STI571-induced cell migration is not fully understood, it appears that inhibition of Abl kinase activity promotes increased tyrosine phosphorylation of Met, which is the receptor for HGF. Met phosphorylation was associated with increased activity of downstream targets, including ERK and Akt, and the response was specific to the Met receptor since there were no changes in phosphorylation of the PDGF receptor in response to STI571.[28] Localization studies revealed that Met and Abl are enriched in the leading pseudopodium of migrating cells, suggesting that their spatial regulation contributes to lamellipodia formation.[28] Although the mechanism is not defined, the results suggest that enhanced phosphorylation of Met promotes increased activation of ERK and Akt at the front of the migrating cell to regulate lamellipodial dynamics. ERK and the PI3K/Akt pathways have been shown to regulate cell migration[2,35,36] and ERK activity is highly localized to the leading pseudopodium where it regulates the actin-myosin system.[35]

Abl activation may also be part of a negative feedback loop that controls the level of Met and Crk phosphorylation and their downstream signals to suppress cell migration.[37] Consistent with this idea, HGF-induced Abl activity is significantly enhanced in spreading cells that are actively engaging the ECM.[28] Crk also transduces signals downstream of the Met receptor that are important for HGF-directed epithelial migration. Together these findings suggest that the Met receptor, integrins, and CAS/Crk cooperate to regulate cell spreading and migration. If this is the case, then inactivation of negative feedback signals from Abl would promote increased CAS/Crk/Rac signaling leading to increased lamellipodium protrusion as discussed above. This in turn would facilitate integrin ligation and membrane spreading over the ECM and enhanced Met activation through cross-talk with integrin receptors. Growth factor receptors are known to associate with integrins and cosignal to the interior of the cell.[38] While CAS/Crk coupling can regulate lamellipodium dynamics, the picture is likely to be more complex than this, as Crk can also couple to the focal adhesion protein paxillin which regulates HGF-induced cell migration through interactions with GIT2/PKL, β-PIX/Cool and Rac.[39] It will be interesting in the future to further define the role of Abl in regulation of Met-induced cell migration and identify the key effectors in this process.

Abl Regulation of PDGF-BB-Induced Cell Migration

In contrast to HGF, Abl activity increases chemotaxis motility following activation of the platelet-derived growth factor receptor (PDGFR) with PDGF-BB.[6,34,40] In an elegant set of experiments it was shown that full Abl and Arg activation by the PDGFR involves a molecular interplay with Src kinase, PLC-γ1, and the PDGFR.[6,34,40] In this system, PDGFR activation decreases the intracellular level of phosphatidyl inositol bisphosphate (PIP2) through PLC-γ1 hydrolysis or by dephosphorylation of PIP2 by inositol polyphosphate 5-phosphatase. It is thought that the decrease in cellular PIP2 levels at specific sites in the membrane relieves the auto-inhibited state of Abl/Arg, the first step in the kinase activation cascade. However, full activation of Abl and Arg requires phosphorylation at several sites, including tyrosine 412 in Abl and 439 in Arg, which are in the activation "loop" of the kinase domain.[40] Although the functional consequences are not fully defined, Abl/Arg and the PDGFR can form a molecular scaffold where they undergo reciprocal phosphorylation.[40] This could provide additional sites for SH2 docking proteins that couple this complex to downstream effectors that mediate cell function or directly regulate tyrosine kinase activity and substrate phosphorylation. Interestingly, while it is not yet known how the PDGFR/Abl/Arg scaffold couples to the migration machinery of the cell, it appears that Abl is the primary player that mediates PDGF-mediated chemotaxis since only Abl and not Arg reconstituted *abl-/-arg-/-* MEF cells showed increased migration in the presence of PDGF-BB.[40] It appears then that Abl kinase activation plays a positive role downstream of the PDGFR to mediate cell migration.

The reasons why Abl negatively regulates HGF and Crk-mediated cell migration but positively regulates PDGFR-directed migration are not known. The simplest scenario is that PDGFR-mediated migration is a separate signaling pathway, which is uncoupled from the negative feedback constraints provided by integrin-mediated signaling events. In this case, Abl may provide an early signaling cue at the membrane that establishes the direction of cell migration in response to a gradient of PDGF-BB. In support of this notion, PLC-γ1 has been shown to play an important role in mediating directional cell migration in response to EGF.[41] On the other hand, the negative feedback function of Abl may be a downstream step in the migration process that operates exclusively to monitor integrin-mediated CAS/Crk coupling after the cell is stimulated to migrate with PDGF-BB. This is consistent with the idea that Abl regulation of CAS/Crk coupling is important for maintaining proper lamellipodial extension and focal adhesion turnover, which is downstream of chemokine gradient sensing mechanisms that initiate and direct cell movement. It will be important to determine the interplay of PDGFR, Abl, and CAS/Crk signaling in the future and how the PDGFR/Abl scaffold couples to the migration machinery of the cell.

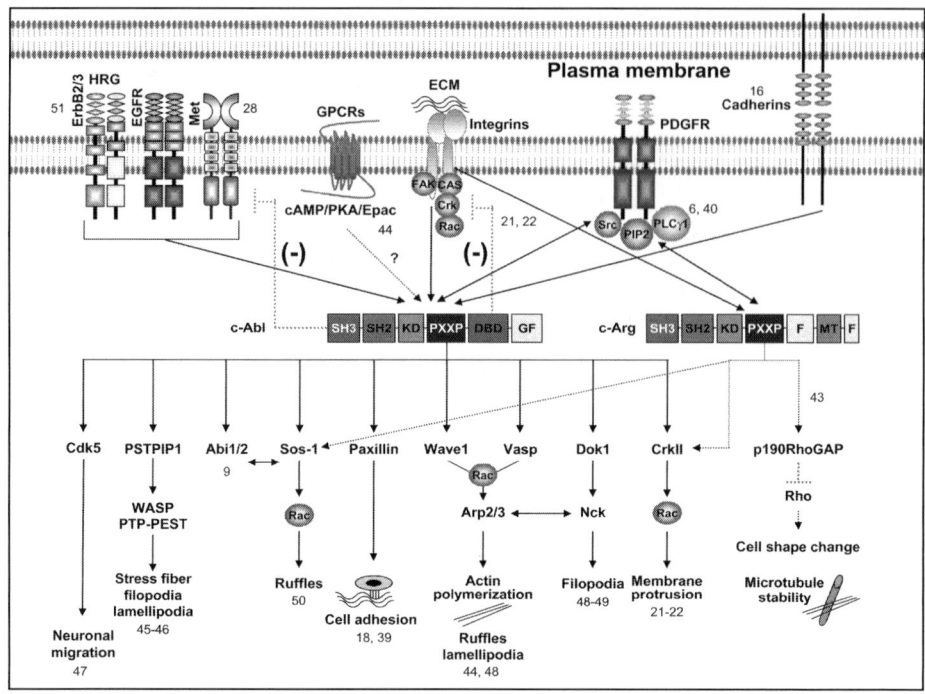

Figure 3. Functional connection map of Abl kinases showing prospective upstream and downstream effectors that mediate cytoskeletal remodeling during cell migration. See references in schematic for details. Heregulin (HRG), EGF receptor family members ErbB2 and ErbB3 (ErbB2/3), hepatocyte growth factor receptor (Met), G-protein coupled receptors (GPCR), protein kinase A (PKA), Exchange factor directly activated by cAMP (Epac), α and β integrins receptors (αβ), extracellular matrix proteins (ECM), Focal adhesion kinase (FAK), platelet-derived growth factor receptor (PDGFR), Crk-associated substrate (CAS), CT10 regulator of kinase (Crk), Phospholipase C gamma one (PLC-γ1), Abelson tyrosine kinase (Abl), Abl-related gene (Arg), Abl interactor 1, 2 (Abi1/2), vasodilator-stimulated phosphoprotein (VASP), Wiskott-Aldrich syndrome protein (WASP), phosphatidyl inositol 4,5-bisphosphate (PIP2).

Future Prospects

While we have presented current evidence for the role of Abl/Arg in cell migration, future work will undoubtedly uncover additional Abl-mediated mechanisms that control cell migration. For example, Abl and Arg associate with and transmit signals to numerous F-actin structures and the microtubule (MT) network, giving rise to microspikes, filopodia, ruffle formation, neurite extension, and synaptic junctions (Fig. 3).[5,7-10,12,15,16] Arg can directly crosslink MT and F-actin where it may serve to insert MT into F-actin protrusive structures in the lamellipodium.[9,15,42] Although this article can not give detailed consideration to this important topic due to space limitations, these direct interactions are likely to assemble key signaling scaffolds that mediate remodeling of the actin cytoskeleton and contribute to morphogenic processes necessary for cell movement. Indeed, a significant amount of evidence indicates that Abl/Arg regulate various effector proteins known to facilitate actin dynamics, including the Rho family of GTPases, Rac and Rho (Fig. 3).[5,7-9,12,15,16,18,22,23,32,43-51] Thus, Abl and Arg are in prime position to serve as cytoskeletal sensors that couple signals emanating from the extracellular environment to the migration machinery of the cell. The challenge will be to decipher the complex signaling cascades and molecular interactions that regulate Abl/Arg activity and their downstream effectors and determine whether these events contribute to physiologically relevant cell migration processes in vivo. Finally, although much is known about oncogenic Bcr-Abl and Tel-Arg[26,52-54] and their role in human leukemia, little is known about whether c-Abl and c-Arg contribute to cancer progression or other unwanted cell migration processes that contribute to inflammation and tumor induced angiogenesis. However, given the central importance of Abl and Arg as monitors of cell shape and movement it seems that future work should consider these non receptor tyrosine kinases as possible contributors to human health and disease.

Acknowledgements

We thank R. Hanley and M. Holcomb and Drs. Y. Wang, O. Pertz and K. Stoletov for helpful comments on the manuscript. The work in the laboratory of R.L.K. is supported by National Institutes of Health Grants CA97022 and GM68487. This is manuscript number 17071-IMM from The Scripps Research Institute.

References

1. Lauffenburger DA, Horwitz AF. Cell migration: A physically integrated molecular process. Cell 1996; 84(3):359-69.
2. Ridley AJ, Schwartz MA, Burridge K et al. Cell migration: Integrating signals from front to back. Science 2003; 302(5651):1704-9.
3. Cho SY, Klemke RL. Extracellular-regulated kinase activation and CAS/Crk coupling regulate cell migration and suppress apoptosis during invasion of the extracellular matrix. J Cell Biol 2000; 149(1):223-36.
4. Cho SY, Klemke RL. Purification of pseudopodia from polarized cells reveals redistribution and activation of Rac through assembly of a CAS/Crk scaffold. J Cell Biol 2002; 156(4):725-36.
5. Kruh GD, Perego R, Miki T et al. The complete coding sequence of arg defines the Abelson subfamily of cytoplasmic tyrosine kinases. Proc Natl Acad Sci USA 1990; 87(15):5802-6.
6. Plattner R, Irvin BJ, Guo S et al. A new link between the c-Abl tyrosine kinase and phosphoinositide signalling through PLC-gamma1. Nat Cell Biol 2003; 5(4):309-19.
7. Hantschel O, Superti-Furga G. Regulation of the c-Abl and Bcr-Abl tyrosine kinases. Nat Rev Mol Cell Biol 2004; 5(1):33-44.
8. Smith JM, Mayer BJ. Abl: Mechanisms of regulation and activation. Front Biosci 2002; 7:d31-42.
9. Woodring PJ, Hunter T, Wang JY. Regulation of F-actin-dependent processes by the Abl family of tyrosine kinases. J Cell Sci 2003; 116(Pt 13):2613-26.
10. Pendergast AM. The Abl family kinases: Mechanisms of regulation and signaling. Adv Cancer Res 2002; 85:51-100.
11. Tybulewicz VL, Crawford CE, Jackson PK et al. Neonatal lethality and lymphopenia in mice with a homozygous disruption of the c-abl proto-oncogene. Cell 1991; 65(7):1153-63.
12. Koleske AJ, Gifford AM, Scott ML et al. Essential roles for the Abl and Arg tyrosine kinases in neurulation. Neuron 1998; 21(6):1259-72.
13. chwartzberg PL, Stall AM, Hardin JD et al. Mice homozygous for the ablm1 mutation show poor viability and depletion of selected B and T cell populations. Cell 1991; 65(7):1165-75.
14. Hardin JD, Boast S, Schwartzberg PL et al. Abnormal peripheral lymphocyte function in c-abl mutant mice. Cell Immunol 1996; 172(1):100-7.
15. Hernandez SE, Krishnaswami M, Miller AL et al. How do Abl family kinases regulate cell shape and movement? Trends Cell Biol 2004; 14(1):36-44.
16. Lanier LM, Gertler FB. From Abl to actin: Abl tyrosine kinase and associated proteins in growth cone motility. Curr Opin Neurobiol 2000; 10(1):80-7.
17. Lewis JM, Baskaran R, Taagepera S et al. Integrin regulation of c-Abl tyrosine kinase activity and cytoplasmic-nuclear transport. Proc Natl Acad Sci USA 1996; 93(26):15174-9.
18. Lewis JM, Schwartz MA. Integrins regulate the association and phosphorylation of paxillin by c-Abl. J Biol Chem 1998; 273(23):14225-30.
19. Ting AY, Kain KH, Klemke RL et al. Genetically encoded fluorescent reporters of protein tyrosine kinase activities in living cells. Proc Natl Acad Sci USA 2001; 98(26):15003-8.
20. Klemke RL, Leng J, Molander R et al. CAS/Crk coupling serves as a "molecular switch" for induction of cell migration. J Cell Biol 1998; 140(4):961-72.
21. Chodniewicz D, Klemke RL. Regulation of integrin-mediated cellular responses through assembly of a CAS/Crk scaffold. Biochim Biophys Acta 2004; 1692(2-3):63-76.
22. Kain KH, Klemke RL. Inhibition of cell migration by Abl family tyrosine kinases through uncoupling of Crk-CAS complexes. J Biol Chem 2001; 276(19):16185-92.
23. Feller SM, Knudsen B, Hanafusa H. c-Abl kinase regulates the protein binding activity of c-Crk. EMBO J 1994; 13(10):2341-51.
24. Kain KH, Gooch S, Klemke RL. Cytoplasmic c-Abl provides a molecular 'Rheostat' controlling carcinoma cell survival and invasion. Oncogene 2003; 22(38):6071-80.

25. Wang B, Mysliwiec T, Feller SM et al. Proline-rich sequences mediate the interaction of the Arg protein tyrosine kinase with Crk. Oncogene 1996; 13(7):1379-85.
26. Druker BJ. Imatinib as a paradigm of targeted therapies. Adv Cancer Res 2004; 91:1-30.
27. Feller SM. Crk family adaptors-signalling complex formation and biological roles. Oncogene 2001; 20(44):6348-71.
28. Frasca F, Vigneri P, Vella V et al. Tyrosine kinase inhibitor STI571 enhances thyroid cancer cell motile response to Hepatocyte Growth Factor. Oncogene 2001; 20(29):3845-56.
29. onaldson LW, Gish G, Pawson T et al. Structure of a regulatory complex involving the Abl SH3 domain, the Crk SH2 domain, and a Crk-derived phosphopeptide. Proc Natl Acad Sci USA 2002; 99(22):14053-8.
30. Pawson T, Gish GD, Nash P. SH2 domains, interaction modules and cellular wiring. Trends Cell Biol 2001; 11(12):504-11.
31. Abassi YA, Vuori K. Tyrosine 221 in Crk regulates adhesion-dependent membrane localization of Crk and Rac and activation of Rac signaling. EMBO J 2002; 21(17):4571-82.
32. Stradal T, Courtney KD, Rottner K et al. The Abl interactor proteins localize to sites of actin poly-merization at the tips of lamellipodia and filopodia. Curr Biol 2001; 11(11):891-5.
33. Tanis KQ, Veach D, Duewel HS et al. Two distinct phosphorylation pathways have additive effects on Abl family kinase activation. Mol Cell Biol 2003; 23(11):3884-96.
34. Plattner R, Kadlec L, DeMali KA et al. c-Abl is activated by growth factors and Src family kinases and has a role in the cellular response to PDGF. Genes Dev 1999; 13(18):2400-11.
35. Brahmbhatt AA, Klemke RL. ERK and RhoA differentially regulate pseudopodia growth and retrac-tion during chemotaxis. J Biol Chem 2003; 278(15):13016-25.
36. Brader S, Eccles SA. Phosphoinositide 3-kinase signalling pathways in tumor progression, invasion and angiogenesis. Tumori 2004; 90(1):2-8.
37. Lamorte L, Royal I, Naujokas M et al. Crk adapter proteins promote an epithelial-mesenchymal-like transition and are required for HGF-mediated cell spreading and breakdown of epithelial adherens junctions. Mol Biol Cell 2002; 13(5):1449-61.
38. Eliceiri BP. Integrin and growth factor receptor crosstalk. Circ Res 2001; 89(12):1104-10.
39. Lamorte L, Rodrigues S, Sangwan V et al. Crk associates with a multimolecular Paxillin/GIT2/beta-PIX complex and promotes Rac-dependent relocalization of Paxillin to focal contacts. Mol Biol Cell 2003; 14(7):2818-31.
40. Plattner R, Koleske AJ, Kazlauskas A et al. Bidirectional signaling links the Abelson kinases to the platelet-derived growth factor receptor. Mol Cell Biol 2004; 24(6):2573-83.
41. Chou J, Burke NA, Iwabu A et al. Directional motility induced by epidermal growth factor requires Cdc42. Exp Cell Res 2003; 287(1):47-56.
42. Miller AL, Wang Y, Mooseker MS et al. The Abl-related gene (Arg) requires its F-actin-microtubule cross-linking activity to regulate lamellipodial dynamics during fibroblast adhesion. J Cell Biol 2004; 165(3):407-19.
43. Hernandez SE, Settleman J, Koleske AJ. Adhesion-dependent regulation of p190RhoGAP in the de-veloping brain by the Abl-related gene tyrosine kinase. Curr Biol 2004; 14(8):691-6.
44. Howe AK. Regulation of actin-based cell migration by cAMP/PKA. Biochim Biophys Acta 2004; 1692(2-3):159-74.
45. Cote JF, Chung PL, Theberge JF et al. PSTPIP is a substrate of PTP-PEST and serves as a scaffold guiding PTP-PEST toward a specific dephosphorylation of WASP. J Biol Chem 2002; 277(4):2973-86.
46. Cong F, Spencer S, Cote JF et al. Cytoskeletal protein PSTPIP1 directs the PEST-type protein ty-rosine phosphatase to the c-Abl kinase to mediate Abl dephosphorylation. Mol Cell 2000; 6(6):1413-23.
47. Courtney MJ, Coffey ET. The mechanism of Ara-C-induced apoptosis of differentiating cerebellar granule neurons. Eur J Neurosci 1999; 11(3):1073-84.
48. Westphal RS, Soderling SH, Alto NM et al. Scar/WAVE-1, a Wiskott-Aldrich syndrome protein, assembles an actin-associated multi-kinase scaffold. EMBO J 2000; 19(17):4589-600.
49. Woodring PJ, Meisenhelder J, Johnson SA et al. c-Abl phosphorylates Dok1 to promote filopodia during cell spreading. J Cell Biol 2004; 165(4):493-503.
50. Sini P, Cannas A, Koleske AJ et al. Abl-dependent tyrosine phosphorylation of Sos-1 mediates growth-factor-induced Rac activation. Nat Cell Biol 2004; 6(3):268-74.
51. Zrihan-Licht S, Avraham S, Jiang S et al. Coupling of RAFTK/Pyk2 kinase with c-Abl and their role in the migration of breast cancer cells. Int J Oncol 2004; 24(1):153-9.
52. Wong S, Witte ON. The BCR-ABL story: Bench to bedside and back. Annu Rev Immunol 2004; 22:247-306.
53. Saglio G, Cilloni D. Abl: The prototype of oncogenic fusion proteins. Cell Mol Life Sci 2004; 61(23):2897-911.
54. Van Etten RA. Mechanisms of transformation by the BCR-ABL oncogene: New perspectives in the post-imatinib era. Leuk Res 2004; 28(Suppl 1):S21-8.

CHAPTER 7

Oncogenic Forms of ABL Family Kinases

Ruibao Ren*

Abstract

Disruption of the auto-inhibitory structure of ABL and ARG activates their kinase activity and oncogenic potential. The oncogenic forms of ABL family kinases, v-ABL, BCR-ABL, TEL-ABL, NUP214-ABL, EML1-ABL, and TEL-ARG, are implicated in a variety of hematological malignancies. The tyrosine kinase activity of all these oncoproteins is essential for the neoplastic transformation, yet additional activities, particularly those of the fusion partners of the ABL kinases, play important roles in determining the lineage and severity of the neoplastic transformation. A better understanding of the mechanism by which the oncogenic forms of ABL family kinases act in leukemogenesis will help to advance therapies for related human leukemias, as well as to understand the mechanism of leukemogenesis and hematopoiesis in general.

Introduction

ABL (also known as ABL1) is a nonreceptor protein tyrosine kinase (Fig. 1) that is expressed in most tissues. Two isoforms of ABL (designated human 1a and 1b in humans or I and IV in mice) are generated by alternative splicing of the first exon; one of them (1b/IV) contains a myristoylation site. Aside from the alternative spliced sequences, the amino (N)-terminal half of ABL contains tandem SRC Homology 3 (SH3), SH2 and the tyrosine kinase domains. These domains can assemble into an auto-inhibitory structure, in which the SH3 and SH2 domains function as a "clamp" (SH2 domain abuts the large lobe of the tyrosine kinase domain; the SH3 domain, the small lobe) that holds the kinase domain in the "off" state.[1,2] In ABL 1b/IV, the N-myristoyl group at the extreme end of the N-terminal segment also binds the large lobe of the tyrosine kinase domain and function as a "latch" that keeps the SH3-SH2 clamp in place.[1,3]

In its carboxyl (C)-terminal region, ABL contains four PXXP SH3 binding motifs, three nuclear localization signals (NLSs), one nuclear exporting signal (NES), a DNA binding domain, and an actin-binding domain [composed of both monomeric (G)- and filamentous (F)-actin binding domains]. The ABL protein is distributed in both nucleus and cytoplasm of cells and can shuttle between the two compartments.[4] It plays an important role in transducing signals from cell surface growth factor and adhesion receptors to regulate cytoskeletal structure (reviewed in refs. 4, 5). Mice with a homozygous disruption of the *ABL* gene—either through a null mutation or a deletion of the C-terminal 1/3 of the protein—are variably affected but some display increased perinatal mortality, runtedness, lymphopenia, osteoporosis and abnormal head and eye development.[6,7]

*Corresponding Author: Ruibao Ren—Rosenstiel Basic Medical Sciences Research Center, MS029, Department of Biology, Brandeis University, 415 South Street, Waltham, Massachusetts 02454, U.S.A. Email: ren@brandeis.edu

Abl Family Kinases in Development and Disease, edited by Anthony Koleske.
©2006 Landes Bioscience and Springer Science+Business Media.

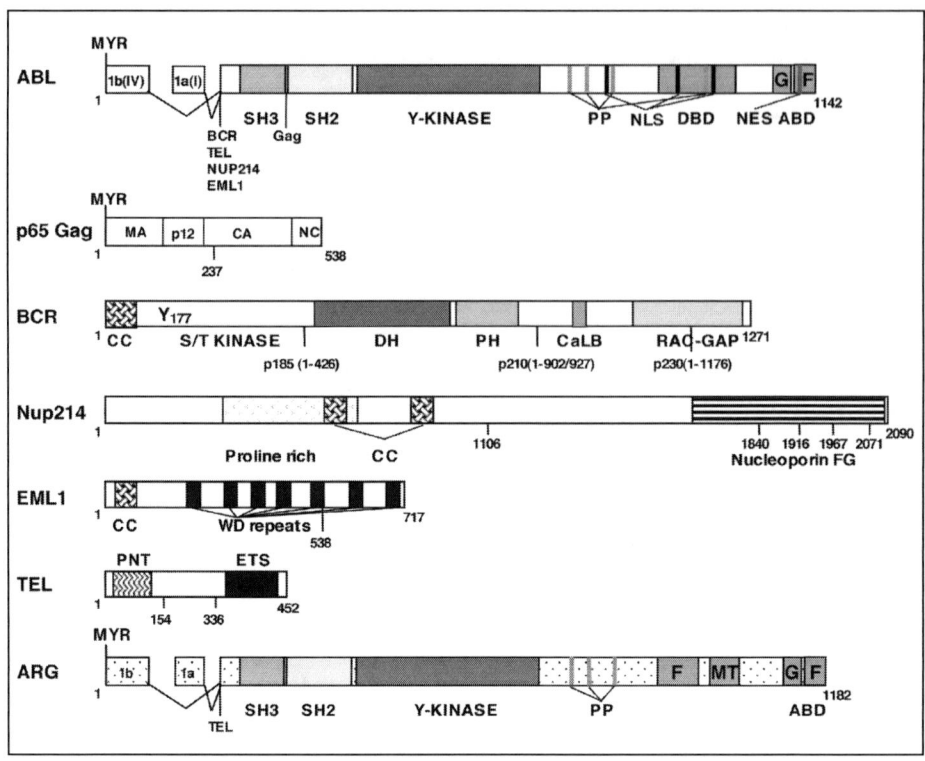

Figure 1. Schematic representation of the ABL, ARG, BCR, TEL, NUP214, EML1 and Mo-MuLV Gag proteins. The domains/motifs that make up ABL, ARG, BCR, TEL, NUP214, and EML1 proteins are shown. Myr: myristoylation site; 1a and 1b: alternative first exon-encoding sequences of the human ABL or ARG; I and IV: alternative first exon-encoding sequences of the murine ABL; SH3 and SH2: SRC-homology 3 and 2 domains; Y-kinase: protein tyrosine kinase; PP: PXXP-containing SH3 binding motifs; NLS: nuclear localization signals; NES: nuclear exporting signal; DBD: DNA binding domain; ABD: actin binding domain [composed of both G- and F-actin binding domains (G and F, respectively)]; CC: Coiled-coil domain; Y177: a critical part of the GRB2-SH2 binding site upon phosphorylation; S/T kinase: serine and threonine kinase; DH: DBL-homo-logy domain; PH: pleckstrin homology domain; CaLB: putative calcium-dependent lipid binding site; RAC-GAP: RAC guanosine triphosphatase-activating protein domain; FG: Phenylalanine/Glycine-rich repeat; PNT: pointed domain; ETS: ETS domain; MT: microtubule-binding domain. The p65 Gag of Mo-MuLV is normally processed proteolytically into matrix (MA), p12, capsid (CA) and nucleocapsid (NC) proteins. Although the Gag MA, p12 and CA cleavage sites are contained within v-ABL, the protein is not cleaved. The fusion points that generate v-ABL, BCR-ABL, NUP214-ABL, EML1-ABL, TEL-ABL and TEL-ARG within each corresponding protein are indicated.

ARG (ABL-related gene, also known as ABL2) is also widely expressed, but is localized primarily in cytoplasm of cells (reviewed in ref. 5). ARG also consists of two isoforms (1a and 1b) (Fig. 1). The SH3, SH2 and tyrosine kinase domains of ABL and ARG are highly homologous (share 89%, 90% and 93% identity, respectively), so the ARG kinase may be auto-regulated in the same way as ABL does. The PXXP SH3 binding motifs and actin binding domain are also highly conserved in ABL and ARG. ARG lacks nuclear localization signals and the DNA binding domain seen in ABL, but contains a second F-actin binding domain and a microtubule-binding domain in its C-terminal region. Consistent with their conserved structure, ABL and ARG share many cellular functions.[8] Although ARG-deficient mice develop normally, mice deficient for both ABL and ARG suffer from defects in neurulation and die

Table 1. **ABL** *family oncogenes and the diseases they are implicated in*

ABL Family Oncogenes	Diseases Implicated In
v-ABL	Murine pre-B cell leukemia/lymphoma
BCR-ABL	*95% human CML*, 20% adult and 2-5% pediatric B-ALL***
TEL-ABL	Rare cases of human CML and B-ALL
TEL-ARG	Rare cases of human AML***
NUP214-ABL	6% human T-ALL****
EML1-ABL	Rare cases of human T-ALL

* Chronic myelogenous leukemia; ** Acute B-lymphoblastic leukemia; *** Acute myelogenous leukemia; **** Acute T-lymphoblastic leukemia

before embryonic day 11. The *ABL⁻/⁻ARG⁻/⁻* neuroepithelial cells display gross alterations in their actin cytoskeleton.

Disruption of the auto-inhibitory structure of ABL and ARG activates their kinase activity and oncogenic potential. The first *ABL* oncogene to be identified was *v-ABL*, the oncogenic element in the Abelson murine leukemia virus (reviewed in see Ref. 9). Subsequently, several naturally occurring oncogenic forms of the ABL family kinases have been identified in humans, including BCR-ABL, TEL-ABL, TEL-ARG, NUP214-ABL, and EML1-ABL.[10-21] These fusion genes are implicated in a variety of human hematological malignancies (Table 1). The mechanisms by which the various oncogenic forms of ABL family kinase cause leukemia are described below.

v-ABL

Abelson murine leukemia virus (A-MuLV) is an acute transforming retrovirus isolated in studying Molony murine leukemia virus (Mo-MuLV, a weekly oncogenic virus)-induced tumors in athymic mice (reviewed in see Ref. 22). A-MuLV causes preB cell leukemia/lymphoma when introduced into newborn mice by intraperitoneal or intravenous injection. In addition, A-MuLV was later found to be able to transform several immortalized murine fibroblast cell lines, primary preB cells and some cytokine-dependent hematopoietic cell lines in vitro.

The oncogenic element of A-MuLV, *v-ABL*, was created by a recombination event that fused viral *gag* sequences (comprised of 271 codons of Mo-MuLV sequences including those encoding matrix protein, p12^{gag}, and 21 amino acids from capsid protein) to murine cellular *ABL* sequences, deleting the ABL N-terminal- and SH3 domain-encoding sequences in the process.[9] Besides the activation of the ABL kinase activity by deletion of the ABL N-termini and SH3 domain, the fusion of Gag sequences to ABL and accumulation of additional mutations within ABL sequence also contribute to the strong oncogenic potential of v-ABL.[9]

Critical Domains and Downstream Signaling Pathways of v-ABL in Transformation

In addition to the ABL tyrosine kinase domain, which is essential for v-ABL transformation both in vitro and in vivo, other regions of the protein are also important for transformation. These include the ABL SH2 domain, C-terminal region and the Gag myristoylation site (Fig. 2). The ABL SH2 domain can mediate both phosphotyrosine-dependent and phosphotyrosine-independent protein-protein interactions.[23,24] One protein that the ABL SH2 domain interacts with is the adaptor molecule SHC.[24] Upon tyrosine phosphorylation, SHC recruits SOS through the SH2- and SH3-containing adaptor protein GRB2 (reviewed in ref. 25). SOS is a guanine nucleotide exchange factor (GEF) for RAS and activates RAS when recruited to the plasma membrane. RAS proteins (H-, N-, and K-RAS4A and 4B) are small

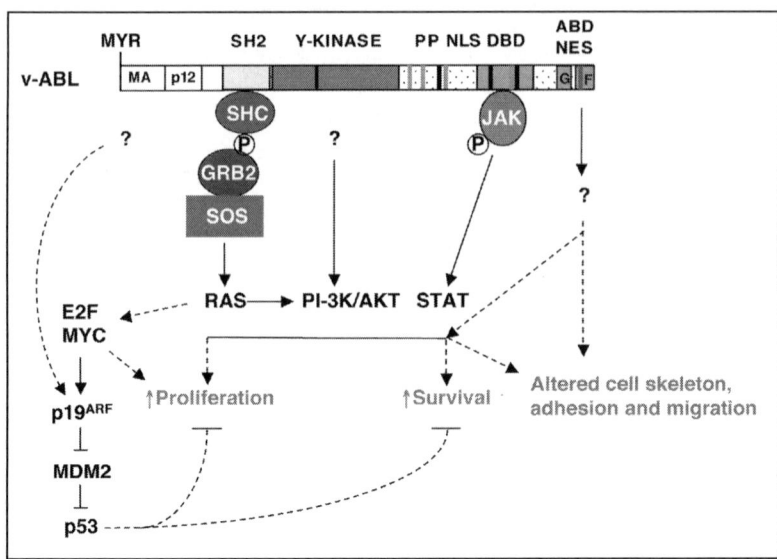

Figure 2. Leukemogenic pathways downstream of v-ABL. v-ABL activates the RAS and STAT signaling pathways through direct interactions with SHC and JAK, respectively. v-ABL may also interact with PI3-K directly. In addition to activating pathways that promote cell proliferation and survival, v-ABL also activates the p53 pathway, which inhibits the cell cycle and induces apoptosis. Inactivation of the p53 pathway is required for v-ABL leukemogenesis. Solid lines indicate direct activation or inhibition. Broken lines indicate multiple steps. Question marks indicate unknown targets or mechanisms. Abbreviations used for Gag sequences and domains of ABL are the same as in Figure 1.

GTPases that act as molecular switches, transducing signals from activated receptors to downstream effectors to regulate cell proliferation, survival and differentiation.[26] The downstream signaling effectors of RAS include the RAF serine/threonine kinase, phosphatidylinositol 3-kinase (PI3-K), RAL-GEFs, Tiam1 (a RAC GEF) and PLCε, all of which are involved in controlling cell proliferation and/or survival.[27] Aberrant activation of RAS proteins, either by *RAS* mutations or by altering genes that directly or indirectly regulate RAS, is common in both solid tumors and hematological malignancies.[28] RAS activation is required for v-ABL transformation, as v-ABL transformation can be blocked by interfering either with *RAS* gene expression or with RAS function.[29,30]

Truncated v-ABL proteins that lack the C-terminal region retain high transforming capacity in NIH 3T3 fibroblast cells but transform lymphocytes poorly.[9,31,32] The central part of the C-terminal region of the murine, but not human, ABL contains a Janus kinase (JAK) binding site, which is required for the activation of STATs (Signal transducers and activators of transcription).[33] The JAK/STAT pathway is an important signaling pathway downstream of many cytokine receptors.[34] Expression of a constitutively activated mutant of STAT5 (STAT5A 1*6) can render 32D cells, an IL-3-dependent hematopoietic cell line, to grow independently of IL-3 and induce a fatal myeloproliferative disorder (MPD) in a mouse bone marrow transduction and transplantation model.[35,36] Disruption of JAK activation by deleting the JAK binding site of v-ABL or by expressing a dominant negative JAK1 abolishes the ability of v-ABL to confer cytokine-independent growth to hematopoietic cell lines and decreased the ability of v-ABL to transform primary lymphocytes.[33,37] In addition to the central part of the C-terminal region, deletion of the ABL actin-binding domain severely reduces the lymphoid transformation, suggesting that the interaction with cytoskeleton is required for v-ABL transformation in lymphocytes.[38] This interaction may contribute to the RAS activation, since expression of an activated RAS can complement the defect of v-ABL with C-terminal deletion mutations in lymphoid transformation.[39]

The myristoylation site of v-ABL is required for the protein to localize to the inner surface of the plasma membrane and has been shown to be essential for transforming fibroblast cells but dispensable for transforming cytokine-dependent hematopoietic cell lines.[31,40] In comparing the leukemogenic potential of wild type (wt) v-ABL versus a v-ABL myristoylation mutant (v-ABL/myr⁻) using a mouse bone marrow transduction and transplantation model, we found that wt v-ABL rapidly induced an acute B-lymphoblastic leukemia (B-ALL) in mice, while v-ABL/myr⁻ induced a T-cell leukemia/lymphoma with a greatly extended latency (Y. Ji and R. Ren, unpublished). This result demonstrates an important role for the myristoylation site both in lineage and severity of v-ABL leukemogenesis. It is possible that efficient activation of signaling pathways such as RAS and PI3-K requires the membrane localization of v-ABL and that such pathways are required for v-ABL transformation in fibroblast cells and B-lymphocytes, but not for IL-3-dependent hematopoietic cell lines. For the latter, activation of the JAK/STAT pathway may be sufficient for their transformation by v-ABL.

Activation of PI3-K and downstream inactivation of FOXO transcription factors are essential for survival of murine pro/preB cells transformed by v-ABL.[41] v-ABL may activate PI3-K via RAS, as well as by directly recruiting PI-3K through an unknown mechanism.[42] An important reaction that PI3-K catalyzes is the phosphorylation of integral membrane phosphatidylinositol 4,5-phosphate (PIP2) at the 3' position.[43] The product phosphatidylinositol 3,4,5-phosphate (PIP3) provides a membrane-docking site that facilitates the activation of a number of signaling proteins, including AKT and Tiam1. AKT is a serine/threonine kinase, which promotes cell survival and cell cycle progression by regulating the apoptosis machinery (e.g., inactivating the pro-apoptotic Bcl-2 related protein, BAD), the transcriptional response to apoptotic stimuli (e.g., inactivating Forkhead factors and activating MDM2 - an E3 ubiquitin ligase which mediates ubiquitylation and proteasome-dependent degradation of the p53 tumor suppressor protein), and glucose metabolism.[25] The PI3-K/AKT pathway is negatively regulated by the PTEN tumor suppressor, a phosphatase that catalyzes the reaction reverting PIP3 back to PIP2.[44] The frequent loss of the PTEN tumor suppressor[45] and high frequency of mutations in the *PI3KCA* gene,[46] which encodes the p110 catalytic subunit of PI3-K, in human cancers further support the importance of PI3-K in oncogenesis.

MYC expression is increased in v-ABL transformed cells and is required for v-ABL transformation, as demonstrated by a dominant negative form of MYC.[47] MYC is a transcription factor that plays a pivotal role in cell cycle regulation, metabolism, apoptosis, and differentiation.[48] Deregulated MYC expression is implicated in a large number of human solid tumors as well as hematological malignancies. v-ABL seems to upregulate c-MYC expression through E2F that is activated by the RAS signaling pathway, since v-ABL induced c-MYC upregulation requires RAS, RAF and cyclin-dependent kinases (which phosphorylate pRb and release its transcription repression from the E2F target genes).[49]

Requirement for Inactivation of the p53 Pathway in v-ABL Leukemogenesis

Despite the fact that v-ABL activates multiple signaling pathways that promote cell proliferation and survival, the v-ABL-induced lymphoid tumors are not polyclonal, and in vitro the primary v-ABL transformants are stimulated to proliferate but subsequently undergo crisis, a period of erratic growth marked by high levels of apoptosis.[22] These data indicate that preB-cell transformation by A-MuLV is a multistep process. Inactivation of the p53 tumor suppressor pathway, either by mutation of p53 or down-modulation of p19^ARF, which activates p53 by sequestrating MDM2, has been frequently observed in v-ABL-induced lymphoid tumor cells (reviewed in ref. 22) and targeted inactivation of p53 facilitates v-ABL leukemogenesis in mice.[50] In addition, v-ABL transformation in cultured preB cells isolated from either p53 or Ink4a/ARF null mice bypass crisis.[51-53] These data indicate that inactivation of the p53 pathway is important for v-ABL leukemogenesis.

E2F and MYC have each been shown to activate the p53 pathway by upregulating p19[ARF].[54] v-ABL may activate the p53 pathway through E2F and MYC, and/or by other unknown mechanisms. The importance of inactivation of the p53 pathway in v-ABL transformation is further supported by the fact that v-ABL could only transform fibroblast cells with the p19[ARF]/ p53 pathway inactivated.[22,55]

BCR-ABL

BCR-ABL is generated by a t(9; 22)(q34; q11) translocation—a hallmark of human chronic myelogenous leukemia (CML) first discovered as an abnormal, small chromosome, named the Philadelphia (Ph) chromosome.[56] BCR is also a signaling protein that contains multiple modular domains (Fig. 1). The fusion of BCR sequences to ABL increases ABL's tyrosine kinase activity, bringing new regulatory domains/motifs, such as the GRB2 SH2 binding site to ABL (Fig. 3).

Different forms of BCR-ABL proteins with different molecular weights (p185, p210 and p230 BCR-ABL) can be generated in patients, depending on the precise break points and RNA splicing (Fig. 1) (reviewed in ref. 57). The most common form, p210 BCR-ABL, is primarily associated with CML. p185 BCR-ABL, on the other hand, is usually associated with B-ALL and rarely with CML. p230 BCR-ABL appears to be a weaker oncogene and is often

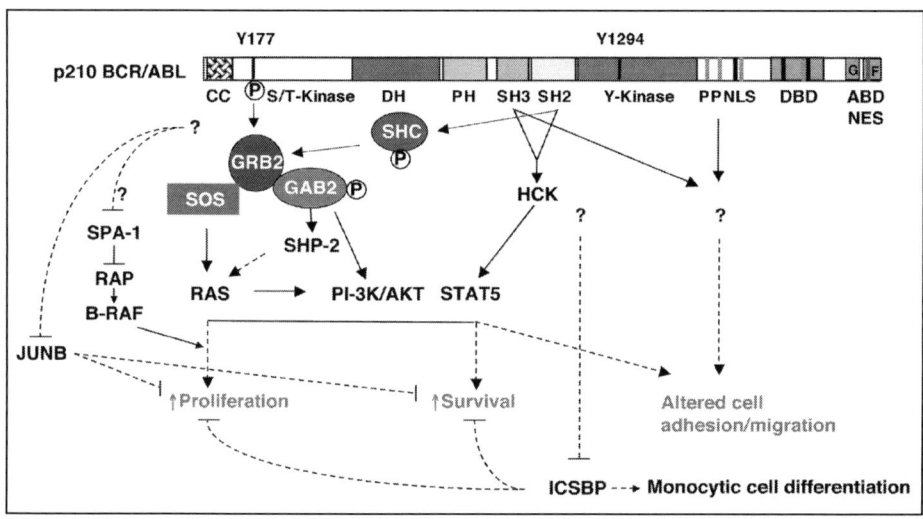

Figure 3. Leukemogenic pathways downstream of BCR-ABL. The activated ABL kinase in BCR-ABL phosphorylates a number of substrates including itself. Phosphorylation at the Y177 residue generates a high affinity binding-site for the GRB2 SH2 domain. GRB2 SH3 domains can bind a number of signaling proteins, including SOS and GAB2. SOS in turn activates RAS. Upon tyrosine phosphorylation, GAB2 recruits PI3-K and SHP2. The ABL SH2 domain can bind SHC, which, upon phosphorylation can also recruit GRB2. The ABL SH3 domain and the SH3 binding sites in the C-terminal region can bind a number of proteins that involve regulations of cell adhesion/migration. The ABL SH3 and SH2 can bind and activate SRC kinase family member, HCK, which can in turn activate STAT5. ICSBP/IRF8 negatively regulates myeloid cell proliferation and survival by inducing monocytic cell differentiation. JUNB inhibits cell proliferation and survival, at least in part by antagonizing the RAS downstream target c-JUN. SPA-1 is a RAP1 GAP, keeping RAP1 inactive. BCR-ABL may promote cell proliferation and survival, at least in part, by activating the RAS, SHP2, PI3-K/AKT and STAT5 signaling pathways and by suppressing ICSBP/IRF8, JUNB and SPA-1. Solid lines indicate direct interactions, activations and/or inhibitions. Broken lines indicate multiple steps. Question marks indicate unknown targets or mechanisms. Abbreviations used for domains of BCR and ABL are the same as in Figure 1.

associated with a less aggressive form of CML. The molecular mechanism of the preferential association of different forms of BCR-ABL with different types of leukemia is not known.

CML and Oncogenic Activity of BCR-ABL

CML is a malignant disease resulting from the neoplastic transformation of a hematopoietic stem cell (HSC).[58] It develops in two phases. The initial chronic phase of CML (CML-CP) is characterized by a massive expansion of the granulocytic cell lineage, even though all hematopoietic lineages can be produced from the CML stem cell. The median duration of CML-CP is 3-4 years. Acquisition of additional genetic abnormalities causes the progression of CML from chronic phase to blast phase (CML-BP), characterized by a block of cell differentiation that results in the presence of 30% or more myeloid or lymphoid blast cells in peripheral blood or bone marrow or the presence of extramedullary infiltrates of blast cells.

Expression of BCR-ABL has been shown to transform established fibroblast cell lines, factor-dependent hematopoietic cell lines, primary mouse bone marrow cells and human CD34+ cells (containing hematopoietic stem cells and hematopoietic progenitor cells) in vitro.[59-61] In mice, expression of BCR-ABL in bone marrow cells, either by retroviral transduction of bone marrow cells followed by transplantation or by transgene-directed inducible expression in HSC, induced a myeloproliferative disorder (MPD) that closely resembles human chronic phase CML.[62-66] These data indicate that BCR-ABL plays a causal role in the pathogenesis of CML-CP.

Imatinib mesylate (Gleevec, previously known as STI571 and CGP 57148), a potent inhibitor of the BCR-ABL tyrosine kinase, has shown a remarkable clinical activity in patients with CML.[67] However, this drug does not completely eradicate BCR-ABL expressing cells from CML patients, and resistance to imatinib emerges.[68-71] Although BCR-ABL remains to be an attractive target for developing CML therapies, identification of additional essential components in the pathogenesis of CML is important for developing strategies for overcoming resistance to imatinib and for eradicating leukemic cells.

Critical Domains and Downstream Signaling Pathways of BCR-ABL in Leukemogenesis

Development of CML is a complex process that involves not only the effects of BCR-ABL, but also the context of its host cells and the rest of the in vivo environment. The mouse bone marrow transduction and bone marrow transplantation (BMT) model provides an effective system for the in vivo analysis of the relative importance of various domains of BCR-ABL and downstream targets of BCR-ABL in leukemogenesis (reviewed in refs. 58, 59). Using this system we have shown that although activation of the ABL tyrosine-kinase activity is essential for BCR-ABL leukemogenesis in mice, other functional domains/motifs are also required for the induction of CML-like disease by BCR-ABL. These include the BCR CC domain, GRB2 binding site at Y177, ABL SH3 and SH2 domains and the PXXP SH3 binding sites.[58]

The N-terminal CC oligomerization domain of BCR is important in activating the ABL kinase and promoting association of BCR-ABL with actin fibers.[72] A mutant BCR-ABL that lacks the BCR-CC domain (ΔCC-BCR-ABL) failed to induce the MPD in mice, and induced a T-cell leukemia/lymphoma only after a long latent period.[73-75] Reactivation of the ABL kinase, by mutating ABL's SH3 domain (deletion or P1013L point mutations), rescued the ability of ΔCC-BCR-ABL to induce the CML-like MPD in mice, albeit with a lesser efficiency.[63,75] These results demonstrate that the BCR-CC domain is essential for the induction of myeloproliferative disorder by BCR-ABL in mice, mainly due to its ability to activate ABL's kinase activity.

The tyrosine 177 is phosphorylated in BCR-ABL, generating a high affinity-binding site for the GRB2 SH2 domain.[76,77] GRB2 binds SOS as well as the scaffolding adapter GAB2 (Fig. 3). Mutation of BCR-ABL's tyrosine-177 residue to phenylalanine (Y177F) largely abolished its ability to bind GRB2, yet did not affect ABL's kinase activity.[76,77] In the BMT model for CML, an Y177F mutant form of BCR-ABL failed to induce the fatal MPD in mice, and induced an acute T-lymphoblastic leukemia (T-ALL) or abdominal T-cell lymphomas after a prolonged latent period.[73,74,78] These results demonstrate that phosphorylation at Y177 is required for the induction of MPD by BCR-ABL.

The GAB2 adapter protein is one of many substrates of BCR-ABL.[79] Consistent with the importance of Y177 in BCR-ABL leukemogenesis, BCR-ABL was unable to confer cytokine-independent growth of primary myeloid cells isolated from GAB2[-/-] mice in vitro,[79] indicating that at least some of the transforming defects associated with the Y177F mutant of BCR-ABL are due to the failure to transmit appropriate signals through GAB2. GAB2 contains binding sites for the SH2 domains of the p85 subunit of PI3-K and SHP2.[79] As mentioned earlier, the PI3-K pathway has been implicated in a wide variety of human cancers.[80] Mutations of the *SHP2* gene (also known as *Ptpn11*) have also been found in approximately 50% of individuals with Noonan syndrome, a common human autosomal dominant birth defect characterized by short stature, facial abnormalities, heart defects and possibly increased risk of leukemia.[81-83] The PI3-K and SHP2 signaling pathways could be required for BCR-ABL leukemogenesis and therefore be effective targets for developing therapeutic interventions for CML.

SHP2 is required for normal activation of the RAS-ERK pathway through which most receptor tyrosine kinases and cytokine receptors signal.[84] The mechanism of the activation of RAS-ERK pathway by SHP2 is not completely known. In addition to SHP2, RAS can be activated directly by BCR-ABL through the GRB2/SOS complex (Fig. 3).[76,85] Recently it was shown that expression of an oncogenic K-RAS in a conditional knock-in mouse strain efficiently induced an MPD resembling human chronic myelomonocytic leukemia (CMML).[86,87] We found that expression of the oncogenic N-RAS (a predominant oncogenic RAS isoform implicated in myeloid malignancies) in mice by bone marrow transduction and transplantation efficiently induced CMML- or acute myelogenous leukemia (AML)-like disease in mice (R. Subrahmanyam, C. Parikh and R. Ren, unpublished). We also found that coexpression of oncogenic N-RAS and Y177F mutant of BCR-ABL could rapidly and efficiently induce CML-like MPD (R. Subrahmanyam, and R. Ren, unpublished). So RAS appears to be a critical downstream target of BCR-ABL.

Another tyrosine phosphorylation site, located in the activation loop of ABL's kinase domain (Y1294 in p210 BCR-ABL), along with ABL's SH2 domain also contribute to the activation of the RAS pathway.[85] As mentioned earlier, ABL's SH2 domain is believed to activate RAS, at least in part, through binding to SHC (Fig. 3). The mechanism of Y1294 in activation of the RAS pathway is not known. We and others have shown that mutation of ABL's SH2 domain reduces the ability of BCR-ABL to induce CML-like MPD in mice.[88,89] An Y1294F point mutation also attenuates leukemogenesis by BCR-ABL.[59,75] These data indicate that both ABL's SH2 domain and phosphorylation site at Y1294 contribute to the overall leukemogenic strength of BCR-ABL.

As mentioned above, the C-terminal region of ABL is required for the proper function of normal ABL and for the lymphoid leukemogenicity of v-ABL.[6,31,32] However, deletion of ABL's actin binding domain or the entire C-terminal region downstream of the ABL kinase domain did not affect the ability of BCR-ABL to induce CML-like MPD in mice,[59,90] indicating that the function of these domains is dispensable in BCR-ABL leukemogenesis.

Certain domains/motifs of BCR-ABL appear to bear overlapping functions. Deletions of both ABL's SH3 domain and the C-terminal proline-rich SH3 binding sites (PP), but not the single mutations, were shown to abrogate BCR-ABL's ability to stimulate spontaneous cell migration on fibronectin-coated surfaces, and to reduce BCR-ABL leukemogenicity in mice.[91] Deletions of both the ABL's SH3 and SH2 domains in BCR-ABL also showed more severe defects in mice than mutating either single domain.[92] It was shown that the ABL SH2 and SH3 domains were required for activation of STAT5, via the SRC family kinase HCK.[93] However, expression of BCR-ABL in bone marrow cells via retroviral transduction still induced

CML-like MPD in STAT5a$^{-/-}$/STAT5b$^{-/-}$ mice.[94] In addition, BCR-ABL expression also induced CML-like MPD in mice lacking the SRC family kinases LYN, HCK and FGR, although it failed to induce a B-ALL in the mutant mice.[95] Further studies are needed for elucidating the mechanism by which the ABL SH2 and SH3 domains affect BCR-ABL leukemogenesis.

Negative Regulators of Myeloid Cell Growth and Their Roles in the Pathogenesis of CML

Gene knock studies have revealed several key negative regulators for myeloid cell growth. Mice with null mutation of Interferon Consensus Sequence Binding Protein (ICSBP)/interferon regulatory factor-8 (IRF-8), JUNB, or SPA-1 (signal-induced proliferation-associated gene-1) developed CML-like diseases.[96-99] ICSBP/IRF-8 controls myeloid cell development by stimulating macrophage differentiation, while inhibiting granulocyte differentiation, in both cases inhibiting cell growth[100] (Fig. 3). JUNB is an antagonist of the RAS downstream target c-JUN and negatively regulates cell proliferation and survival. SPA-1 is a principal RAP1 GTPase-activating protein (GAP) in hematopoietic progenitors.[99] RAP1 is a close member of RAS-family GTPases and can activate the MEK/ERK signaling pathway through activation of B-RAF.[101]

We have found that expression of the ICSBP/IRF-8 protein is significantly decreased in mice with BCR-ABL-induced CML-like disease and forced expression of ICSBP/IRF-8 inhibits the BCR-ABL-induced colony-formation of bone marrow cells in vitro and BCR-ABL-induced CML-like disease in vivo.[102] Downregulation of *ICSBP/IRF-8* transcripts has also been found in CML patients and that this reduction of *ICSBP/IRF-8* transcripts can be reversed by interferon (IFN)-α treatment.[103] In addition, downregulation of JUNB has been observed in CML cells isolated from patients.[104,105] These results indicate that ICSBP/IRF-8 and JUNB are tumor suppressors, and that downregulation of ICSBP/IRF-8 and JUNB is important in the pathogenesis of CML.

It is not clear whether SPA-1 is downregulated in CML, but it was shown that expression of BCR-ABL in a factor-dependent hematopoietic cell line activated RAP1 and B-RAF.[106] These results indicate that in addition to the well-established RAS signaling pathway, BCR/ABL may activate MEK/ERK through the alternative-signaling pathway involving RAP1 and B-RAF (Fig. 3).

Mechanism of CML vs. B-ALL Induced by ABL Oncogenes

The molecular mechanism by which BCR-ABL causes CML vs. B-ALL is not completely known. BCR-ABL can transform B-cell precursors in vitro.[107] It was also shown that BCR-ABL induced a CML-like MPD when bone marrow cells from 5-fluorouracil (5-FU)-treated mice were used, but induced a B-ALL when bone marrow cells from non5-FU-treated mice were used in the BMT experiment (reviewed in ref. 59). The purpose of 5-FU treatment is to eliminate the proliferating hematopoietic precursor cells and to enrich and stimulate HSCs. In addition, several transgenic mouse lines expressing BCR-ABL developed B-ALL.[108-112] It appears that BCR-ABL induces CML and B-ALL by targeting different cells—HSC in the case of CML vs. lymphocyte precursors in the case of B-ALL.

The different diseases that BCR-ABL and v-ABL induce appear to be determined by the intrinsic oncogenic properties of the oncogenes. We have shown that v-ABL has a much lower capacity for stimulating myeloid cell growth than BCR-ABL both in vitro and in vivo, and that v-ABL induces primarily a B-ALL in mice under the same conditions in which BCR-ABL induces a CML-like MPD.[113] Further identifying genetic determinant(s) in BCR-ABL and v-ABL and their downstream targets that control the lineage specific transformation will help to understand the mechanisms in the pathogenesis of myeloid vs. lymphoid malignancies.

TEL-ABL and TEL-ARG

TEL-ABL and Its Leukemogenicity

The *TEL-ABL* fusion gene is the product of a t(9; 12)(q34; p13) translocation found in rare cases of human CML and B-ALL.[11-16] It shares the same fusion point on *ABL* as *BCR-ABL*. TEL (also known as ETV6) is a member of the ETS family of transcription regulators. Two naturally occurring forms of *TEL-ABL* contain either exons 1-4 (type A) or exons 1-5 (type B) of the *TEL* gene, which encodes the N-terminal 1-154 or 1-336 amino acids of TEL respectively. The extra sequences in the type B TEL-ABL (TEL-ABL B) contain a GRB2-SH2 binding site.[114] However, both TEL-ABL proteins contain a pointed (PNT) homology domain (also known as SAM, SPM or HLH domains). Similar to the N-terminal CC domain of BCR-ABL the PNT domain of TEL can mediate oligomerization of the fusion protein, and is required for activation of the ABL kinase, localization of TEL-ABL along actin filaments, and hematopoietic transformation by TEL-ABL.[12,115-118] It has been shown that BCR-ABL and TEL-ABL activate similar signal transduction pathways in hematopoietic cell lines.[119]

In the same BMT model system where BCR-ABL induces a CML-like MPD, both type A and type B TEL-ABL can induce two distinct myeloproliferative diseases — a predominant myeloproliferative small bowel disease (SBD), characterized by a moderate leukocytosis and splenomegaly, extensive myeloid cell infiltration and necrosis in small intestine and endotoxemia, and a CML-like MPD that is similar to that induced by BCR/ABL.[59,118,120] The TEL-ABL associated SBD has not been observed clinically. We have recently found that the high expression of a transgene driven by a retroviral vector can be modulated by internal ribosome entry site (IRES) variants (reviewed in ref. 121). Using both modulated and conventional retroviral vectors, we found that TEL-ABL A induced a CML-like MPD versus a myeloproliferative SBD in mice in a dose-dependent manner. Since TEL-ABL A lacks the GRB2-SH2 binding site, which is required for the induction of MPD by BCR-ABL, further studying the mechanisms by which TEL-ABL causes MPD will help elucidate the common molecular mechanism of CML.

TEL-ARG

The *TEL-ARG* fusion gene is generated by a t(1; 12)(q25; p13) reciprocal translocation found in rare cases of human AML.[17,18] It was shown that expression of TEL-ARG in cytokine dependent Ba/F3 murine hematopoietic cells resulted in prolonged viability and hyper-responsiveness to IL-3.[122,123] The PNT domain in TEL and the tyrosine kinase activity of ARG were essential for both signaling and biological effects. The SH3 domain in ARG was required for hyper-responsiveness to IL-3, but not for prolonged viability. The opposite was true for the SH2 domain of ARG. Mutation of Y314 in TEL, a putative GRB2-binding site, led to reduced viability, and loss of hyper-responsiveness to IL-3. Interestingly, all biological functions were profoundly impaired with deletion of the C-terminus in ARG.[122] The mechanism of TEL-ARG in the pathogenesis of AML remains unknown. It would be interesting to compare the leukemogenic potential of TEL-ARG with other ABL oncogenes using the mouse BMT model.

NUP214-ABL and EML1-ABL

NUP214-ABL

NUP214-ABL is generated by an extrachromosomal (episomal) amplification of ABL in approximately 6% individuals with T-ALL.[19,20] The amplicon is a 500-kb region from chromosome band 9q34, containing the ABL and NUP214 genes. Different from the reciprocal chromosome translocation, the NUP214-ABL fusion is generated by the formation of episome (circularization of the amplified DNA segment). The human *NUP214* (also known as CAN) gene was first identified as a target of t(6; 9)(p23; q34), associated with AML and myelodysplastic syndrome, which results in the expression of a DEK-CAN fusion gene.[124,125] NUP214, is a nuclear pore complex (NPC) protein that contains multiple Phenylalanine/ Glycine (FG)-rich repeat motifs (Fig. 1). It interacts at the NPC with at least two other proteins, the nucleoporin NUP88 and hCRM1 (exportin 1), which was shown to function as a nuclear export receptor.[126]

Like other ABL oncogenes, NUP214-ABL is constitutively phosphorylated and is sensitive to the tyrosine kinase inhibitor imatinib. The coiled-coil domains of NUP214 may mediate oligomerization of the fusion protein and contribute to the activation of the ABL tyrosine kinase.

EML1-ABL

EML1-ABL was found in a T-ALL patient with a cryptic t(9; 14)(q34; q32) transloca-tion.[21] EML1 encodes a protein with high similarity to the echinoderm microtubule-associated protein, a WD-repeat protein associated with the mitotic spindle apparatus.[127] EML1-ABL is also a constitutively phosphorylated tyrosine kinase and transforms Ba/F3 cells to growth fac-tor- independent growth through activation of survival and proliferation pathways, including ERK1/2, STAT5 and LYN kinase.[21] Interestingly, EML1 also contains a N-terminal coiled-coil domain (Fig. 1) and deletion of this CC domain abrogates the transforming properties of EML1-ABL.[21]

The mechanism of NUP214-ABL and EML1-ABL in the pathogenesis of T-ALL is not known. It is interesting that several BCR-ABL mutants, such as Y177F and ΔCC, as well as the v-ABL/myr⁻ mutant described above, induce T-ALL in mice with a long latency. The long incubation time for the development of T-cell tumors suggests that secondary mutations may be involved. It is possible that NUP214-ABL and EML1-ABL are weaker oncogenes compared to BCR-ABL and that they are insufficient to induce CML and even B-ALL. This hypothesis can be tested in mice using the BMT model system. In addition, consistent with a multistep pathogenesis of T-ALL, NUP214-ABL and EML1-ABL were found to be associated with in-creased HOX gene expression and deletion of CDKN2A (p16).

Future Directions

The tyrosine kinase activity is essential for the neoplastic transformation by the ABL family oncoproteins. However, the ABL kinase inhibitors, such as imatinib, appear to be a suppressive rather than curative therapeutics. Other functions of the ABL family oncoproteins along with the biology of host cells and tumor microenvironment are likely to contribute to the disease development and maintenance.[58] Although the ABL kinase remains to be an attractive target for developing leukemia therapies, further elucidating the mechanisms of the ABL family oncoproteins in the neoplastic transformation and identifying additional essential components in leukemogenesis are important for developing strategies to overcome resistance to ABL ki-nase inhibitors and for eradicating leukemic cells.

It is interesting that different ABL family oncoproteins are associated with different types of hematopoietic malignancies. It is evident that the type of hematopoietic malignancies induced by an *ABL* family oncogene can be determined by the types of hematopoietic cells the oncogene is targeted into and/or by the intrinsic oncogenic properties of the particular *ABL* family oncogene. It is important to identify the origin cell of various hematopoietic malignancies involving the *ABL* family oncogenes, to compare the in vivo leukemogencity of various *ABL* family oncogenes in animal models and to identify specific signaling pathways downstream of various *ABL* family oncogenes. Such studies will help to develop therapies eradicating leukemia initiating/stem cells as well as to understand the mechanism of lineage specific malignancies.

Addition genetic and epigenetic abnormalities are involved in the disease progression of CML[58] as well as in the pathogenesis of the acute leukemias (described above) involving the *ABL* family oncogenes. Further identifying such abnormalities and studying the mechanism by which the *ABL* family oncogenes cooperate with such abnormalities in leukemogenesis are clearly important for elucidating the mechanism of leukemogenesis and for advancing thera-pies for human leukemias.

References

1. Nagar B, Hantschel O, Young MA et al. Structural basis for the autoinhibition of c-Abl tyrosine kinase. Cell 2003; 112:859-871.
2. Harrison SC. Variation on an Src-like theme. Cell 2003; 112:737-740.
3. Hantschel O, Nagar B, Guettler S et al. A myristoyl/phosphotyrosine switch regulates c-Abl. Cell 2003; 112:845-857.
4. Woodring PJ, Hunter T, Wang JY. Regulation of F-actin-dependent processes by the Abl family of tyrosine kinases. J Cell Sci 2003; 116:2613-2626.
5. Hernandez SE, Krishnaswami M, Miller AL et al. How do Abl family kinases regulate cell shape and movement? Trends Cell Biol 2004; 14:36-44.
6. Schwartzberg PL, Stall AM, Hardin JD et al. Mice homozygous for the ablm1 mutation show poor viaility and depletion of selected B and T cell populations. Cell 1991; 65:1165-1175.
7. Tybulewicz VLJ, Crawford CE, Jackson PK et al. Neonatal lethality and lymphopenia in mice with a homozygous disruption of the c-abl proto-oncogene. Cell 1991; 65:1153-1163.
8. Koleske AJ, Gifford AM, Scott ML et al. Essential roles for the Abl and Arg tyrosine kinases in neurulation. Neuron 1998; 21:1259-1272.
9. Rosenberg N, Witte ON. The viral and cellular forms of the abelson (abl) oncogene. Advances in virus research. Vol 35. New York: Academic Press Inc., 1988:39-81.
10. Melo JV. The diversity of Bcr-Abl fusion proteins and their relationship to leukemia phenotype. Blood 1996; 88:2375-2384.
11. Papadopoulos P, Ridge SA, boucher CA et al. The novel activation of Abl by Fusion to an ets-related gene, TEL. Cancer Research 1995; 55:34-38.
12. Golub TR, Goga A, Barker GF et al. Oligomerization of the ABL tyrosine kinase by the Ets protein TEL in human leukemia. Mol Cell Biol 1996; 16:4107-4116.
13. Brunel V, Sainty D, Carbuccia N et al. A TEL/ABL fusion gene on chromosome 12p13 in a case of Ph-, BCR- atypical CML. Leukemia 1996; 10:2003.
14. Andreasson P, Johansson B, Carlsson M et al. BCR/ABL-negative chronic myeloid leukemia with ETV6/ABL fusion. Genes Chromosomes Cancer 1997; 20:299-304.
15. Van Limbergen H, Beverloo HB, van Drunen E et al. Molecular cytogenetic and clinical findings in ETV6/ABL1-positive leukemia. Genes Chromosomes Cancer 2001; 30:274-282.
16. Keung YK, Beaty M, Steward W et al. Chronic myelocytic leukemia with eosinophilia, t(9; 12)(q34; p13), and ETV6-ABL gene rearrangement: Case report and review of the literature. Cancer Genet Cytogenet 2002; 138:139-142.
17. Cazzaniga G, Tosi S, Aloisi A et al. The tyrosine kinase abl-related gene ARG is fused to ETV6 in an AML-M4Eo patient with a t(1; 12)(q25; p13): Molecular cloning of both reciprocal transcripts. Blood 1999; 94:4370-4373.
18. Iijima Y, Ito T, Oikawa T et al. A new ETV6/TEL partner gene, ARG (ABL-related gene or ABL2), identified in an AML-M3 cell line with a t(1; 12)(q25; p13) translocation. Blood 2000; 95:2126-2131.
19. Graux C, Cools J, Melotte C et al. Fusion of NUP214 to ABL1 on amplified episomes in T-cell acute lymphoblastic leukemia. Nat Genet 2004; 36:1084-1089.
20. Ballerini P, Busson M, Fasola S et al. NUP214-ABL1 amplification in t(5; 14)/HOX11L2-positive ALL present with several forms and may have a prognostic significance. Leukemia 2005; 19:468-470.
21. De Keersmaecker K, Graux C, Odero MD et al. Fusion of EML1 to ABL1 in T-cell acute lymphoblastic leukemia with cryptic t(9; 14)(q34; q32). Blood 2005.
22. Shore SK, Tantravahi RV, Reddy EP. Transforming pathways activated by the v-Abl tyrosine kinase. Oncogene 2002; 21:8568-8576.
23. Pawson T. Protein modules and signalling networks. Nature 1995; 373:573-580.
24. Raffel GD, Parmar K, Rosenberg N. In vivo association of v-Abl with Shc mediated by a nonphosphotyrosine- dependent SH2 interaction. J Biol Chem 1996; 271:4640-4645.
25. Downward J. Targeting RAS signalling pathways in cancer therapy. Nat Rev Cancer 2003; 3:11-22.
26. Ulku AS, Der CJ. Ras signaling, deregulation of gene expression and oncogenesis. Cancer Treat Res 2003; 115:189-208.
27. Repasky GA, Chenette EJ, Der CJ. Renewing the conspiracy theory debate: Does Raf function alone to mediate Ras oncogenesis? Trends Cell Biol 2004; 14:639-647.
28. Bos JL. RAS oncogenes in human cancer: A review [published erratum appears in Cancer Res 1990 Feb 15; 50(4):1352]. Cancer Res 1989; 49:4682-4689.
29. Sawyers CL, McLaughlin J, Witte ON. Genetic requirement for Ras in the transformation of fibroblasts and hematopoietic cells by the Bcr-Abl oncogene. J Exp Med 1995; 181:307-313.
30. Stacey DW, Roudebush M, Day R et al. Dominant inhibitory Ras mutants demonstrate the requirement for Ras activity in the action of tyrosine kinase oncogenes. Oncogene 1991; 6:2297-2304.

31. Prywes R, Foulkes JG, Rosenberg N et al. Sequences of the A-MuLV protein needed for fibroblast and lymphoid cell transformation. Cell 1983; 34:569-579.
32. Parmar K, Huebner RC, Rosenberg N. Carboxyl-terminal determinants of Abelson protein important for lymphoma induction. J Virol 1991; 65:6478-6485.
33. Danial NN, Losman JA, Lu T et al. Direct interaction of Jak1 and v-Abl is required for v-Abl-induced activation of STATs and proliferation. Mol Cell Biol 1998; 18:6795-6804.
34. Leonard WJ, O'Shea JJ. Jaks and STATs: Biological implications. Annu Rev Immunol 1998; 16:293-322.
35. Onishi M, Nosaka T, Misawa K et al. Identification and characterization of a constitutively active STAT5 mutant that promotes cell proliferation. Mol Cell Biol 1998; 18:3871-3879.
36. Schwaller J, Parganas E, Wang D et al. Stat5 is essential for the myelo- and lymphoproliferative disease induced by TEL/JAK2. Mol Cell 2000; 6:693-704.
37. Limnander A, Danial NN, Rothman PB. v-Abl signaling disrupts SOCS-1 function in transformed preB cells. Mol Cell 2004; 15:329-341.
38. Warren D, Griffin DS, Mainville C et al. The extreme carboxyl terminus of v-Abl is required for lymphoid cell transformation by Abelson virus. J Virol 2003; 77:4617-4625.
39. Parmar K, Rosenberg N. Ras complements the carboxy terminus of v-Abl protein in lymphoid transformation. J Virol 1996; 70:1009-1015.
40. Daley GQ, Van Etten RA, Jackson PK et al. Nonmyrityolated Abl proteins transform a Factor-dependent hematopoietic cell line. Mol Cell Biol 1992; 12:1864-1871.
41. Kharas MG, Deane JA, Wong S et al. Phosphoinositide 3-kinase signaling is essential for ABL oncogene mediated transformation of B lineage cells. Blood 2004.
42. Varticovski L, Daley GQ, Jackson P et al. Activation of phosphtidylinositol 3-kinase in cells expressing abl oncogene variants. Mol Cell Biol 1991; 11:1107-1113.
43. Downward J. PI 3-kinase, Akt and cell survival. Semin Cell Dev Biol 2004; 15:177-182.
44. Leslie NR, Downes CP. PTEN: The down side of PI 3-kinase signalling. Cell Signal 2002; 14:285-295.
45. Simpson L, Parsons R. PTEN: Life as a tumor suppressor. Exp Cell Res 2001; 264:29-41.
46. Samuels Y, Wang Z, Bardelli A et al. High frequency of mutations of the PIK3CA gene in human cancers. Science 2004; 304:554.
47. Sawyers CL, Callahan W, Witte ON. Dominant negative MYC blocks transformation by ABL oncogenes. Cell 1992; 70:901-910.
48. Hoffman B, Amanullah A, Shafarenko M et al. The proto-oncogene c-myc in hematopoietic development and leukemogenesis. Oncogene 2002; 21:3414-3421.
49. Zou X, Rudchenko S, Wong K et al. Induction of c-myc transcription by the v-Abl tyrosine kinase requires Ras, Raf1, and cyclin-dependent kinases. Genes Dev 1997; 11:654-662.
50. Zou X, Cong F, Coutts M et al. p53 deficiency increases transformation by v-Abl and rescues the ability of a C-terminally truncated v-Abl mutant to induce preB lymphoma in vivo. Mol Cell Biol 2000; 20:628-633.
51. Radfar A, Unnikrishnan I, Lee HW et al. p19(Arf) induces p53-dependent apoptosis during abelson virus-mediated preB cell transformation. Proc Natl Acad Sci USA 1998; 95:13194-13199.
52. Unnikrishnan I, Radfar A, Jenab-Wolcott J et al. p53 mediates apoptotic crisis in primary Abelson virus-transformed preB cells. Mol Cell Biol 1999; 19:4825-4831.
53. Unnikrishnan I, Rosenberg N. Absence of p53 complements defects in Abelson murine leukemia virus signaling. J Virol 2003; 77:6208-6215.
54. Matsumura I, Tanaka H, Kanakura Y. E2F1 and c-Myc in cell growth and death. Cell Cycle 2003; 2:333-338.
55. Cong F, Zou X, Hinrichs K et al. Inhibition of v-Abl transformation by p53 and p19ARF. Oncogene 1999; 18:7731-7739.
56. Kurzrock R, Gutterman J, Talpaz M. The molecular genetics of Philadelphia chromosome-positive leukemias. New Engl J Med 1988; 319:990-998.
57. Melo JV. BCR-ABL gene variants. Baillieres Clin Haematol 1997; 10:203-222.
58. Ren R. Mechanisms of BCR-ABL in the pathogenesis of chronic myelogenous leukaemia. Nat Rev Cancer 2005; 5:172-183.
59. Ren R. The molecular mechanism of chronic myelogenous leukemia and its therapeutic implications: Studies in a murine model. Oncogene 2002; 21:8629-8642.
60. Ramaraj P, Singh H, Niu N et al. Effect of mutational inactivation of tyrosine kinase activity on BCR/ABL-induced abnormalities in cell growth and adhesion in human hematopoietic progenitors. Cancer Res 2004; 64:5322-5331.
61. Zhao RC, Jiang Y, Verfaillie CM. A model of human p210(bcr/ABL)-mediated chronic myelogenous leukemia by transduction of primary normal human CD34(+) cells with a BCR/ABL- containing retroviral vector. Blood 2001; 97:2406-2412.

62. Daley GQ, Van Etten RA, Baltimore D. Induction of chronic myelogenous leukemia in mice by the P210bcr/abl gene of the Philadelphia chromosome. Science 1990; 247:824-830.

63. Zhang X, Ren R. Bcr-Abl efficiently induces a myeloproliferative disease and production of excess interleukin-3 and granulocyte-macrophage colony-stimulating factor in mice: A novel model for chronic myelogenous leukemia. Blood 1998; 92:3829-3840.

64. Pear WS, Miller JP, Xu L et al. Efficient and rapid induction of a chronic myelogenous leukemia-like myeloproliferative disease in mice receiving P210 bcr/abl-transduced bone marrow. Blood 1998; 92:3780-3792.

65. Li S, Ilaria RL, Million RP et al. The P190, P210, and P230 forms of the BCR/ABL oncogene induce a similar chronic myeloid leukemia-like syndrome in mice but have different lymphoid leukemogenic activity. J Exp Med 1999; 189:1399-1412.

66. Koschmieder S, Gottgens B, Zhang P et al. Inducible chronic phase of myeloid leukemia with expansion of hematopoietic stem cells in a transgenic model of BCR-ABL leukemogenesis. Blood 2005; 105:324-334.

67. Druker BJ. Inhibition of the Bcr-Abl tyrosine kinase as a therapeutic strategy for CML. Oncogene 2002; 21:8541-8546.

68. Stentoft J, Pallisgaard N, Kjeldsen E et al. Kinetics of BCR-ABL fusion transcript levels in chronic myeloid leukemia patients treated with STI571 measured by quantitative real-time polymerase chain reaction. Eur J Haematol 2001; 67:302-308.

69. Bhatia R, Holtz M, Niu N et al. Persistence of malignant hematopoietic progenitors in chronic myelogenous leukemia patients in complete cytogenetic remission following imatinib mesylate treatment. Blood 2003; 101:4701-4707.

70. Lowenberg B. Minimal residual disease in chronic myeloid leukemia. N Engl J Med 2003; 349:1399-1401.

71. Gorre ME, Sawyers CL. Molecular mechanisms of resistance to STI571 in chronic myeloid leukemia. Curr Opin Hematol 2002; 9:303-307.

72. McWhirter JR, Gaalasso DL, Wang JYJ. A coiled-coil oligomerization domain of Bcr is essential for the transforming function of Bcr-Abl oncoproteins. Mol Cell Biol 1993; 13:7587-7595.

73. Zhang X, Subrahmanyam R, Wong R et al. The NH2-terminal coiled-coil domain and tyrosine 177 play important roles in induction of a myeloproliferative disease in mice by Bcr-Abl. Mol Cell Biol 2001; 21:840-853.

74. He Y, Wertheim JA, Xu L et al. The coiled-coil domain and Tyr177 of bcr are required to induce a murine chronic myelogenous leukemia-like disease by bcr/abl. Blood 2002; 99:2957-2968.

75. Smith KM, Yacobi R, Van Etten RA. Autoinhibition of Bcr-Abl through its SH3 domain. Mol Cell 2003; 12:27-37.

76. Pendergast AM, Quilliam LA, Cripe LD et al. BCR-ABL-induced oncogenesis is mediated by direct interaction with the SH2 domain of the Grb-2 adaptor protein. Cell 1993; 75:175-185.

77. Puil L, Liu J, Gish G et al. Bcr-Abl oncoproteins bind directly to activators of the Ras signalling pathway. EMBO J 1994; 13:764-773.

78. Million RP, Van Etten RA. The Grb2 binding site is required for the induction of chronic myeloid leukemia-like disease in mice by the Bcr/Abl tyrosine kinase. Blood 2000; 96:664-670.

79. Sattler M, Mohi MG, Pride YB et al. Critical role for Gab2 in transformation by BCR/ABL. Cancer Cell 2002; 1:479-492.

80. Vivanco I, Sawyers CL. The phosphatidylinositol 3-Kinase AKT pathway in human cancer. Nat Rev Cancer 2002; 2:489-501.

81. Tartaglia M, Mehler EL, Goldberg R et al. Mutations in PTPN11, encoding the protein tyrosine phosphatase SHP-2, cause Noonan syndrome. Nat Genet 2001; 29:465-468.

82. Tartaglia M, Niemeyer CM, Fragale A et al. Somatic mutations in PTPN11 in juvenile myelomonocytic leukemia, myelodysplastic syndromes and acute myeloid leukemia. Nat Genet 2003; 34:148-150.

83. Kosaki K, Suzuki T, Muroya K et al. PTPN11 (protein-tyrosine phosphatase, nonreceptor-type 11) mutations in seven Japanese patients with Noonan syndrome. J Clin Endocrinol Metab 2002; 87:3529-3533.

84. Neel BG, Gu H, Pao L. The 'Shp'ing news: SH2 domain-containing tyrosine phosphatases in cell signaling. Trends Biochem Sci 2003; 28:284-293.

85. Goga A, McLaughlin J, Afar DE et al. Alternative signals to RAS for hematopoietic transformation by the Bcr-Abl oncogene. Cell 1995; 82:981-988.

86. Braun BS, Tuveson DA, Kong N et al. Somatic activation of oncogenic Kras in hematopoietic cells initiates a rapidly fatal myeloproliferative disorder. Proc Natl Acad Sci USA 2004; 101:597-602.

87. Chan IT, Kutok JL, Williams IR et al. Conditional expression of oncogenic K-ras from its endogenous promoter induces a myeloproliferative disease. J Clin Invest 2004; 113:528-538.

88. Zhang X, Wong R, Hao SX et al. The SH2 domain of bcr-Abl is not required to induce a murine myeloproliferative disease; however, SH2 signaling influences disease latency and phenotype. Blood 2001; 97:277-287.

89. Roumiantsev S, de Aos IE, Varticovski L et al. The src homology 2 domain of Bcr/Abl is required for efficient induction of chronic myeloid leukemia-like disease in mice but not for lymphoid leukemogenesis or activation of phosphatidylinositol 3-kinase. Blood 2001; 97:4-13.

90. Wertheim JA, Perera SA, Hammer DA et al. Localization of BCR-ABL to F-actin regulates cell adhesion but does not attenuate CML development. Blood 2003; 102:2220-2228.

91. Dai Z, Kerzic P, Schroeder WG et al. Deletion of the Src homology 3 domain and C-terminal proline-rich sequences in Bcr-Abl prevents Abl interactor 2 degradation and spontaneous cell migration and impairs leukemogenesis. J Biol Chem 2001; 276:28954-28960.

92. Nieborowska-Skorska M, Hoser G, Kossev P et al. Complementary functions of the antiapoptotic protein A1 and serine/threonine kinase pim-1 in the BCR/ABL-mediated leukemogenesis. Blood 2002; 99:4531-4539.

93. Klejman A, Schreiner SJ, Nieborowska-Skorska M et al. The Src family kinase Hck couples BCR/ABL to STAT5 activation in myeloid leukemia cells. EMBO J 2002; 21:5766-5774.

94. Sexl V, Piekorz R, Moriggl R et al. Stat5a/b contribute to interleukin 7-induced B-cell precursor expansion, but abl- and bcr/abl-induced transformation are independent of stat5. Blood 2000; 96:2277-2283.

95. Hu Y, Liu Y, Pelletier S et al. Requirement of Src kinases Lyn, Hck and Fgr for BCR-ABL1-induced B-lymphoblastic leukemia but not chronic myeloid leukemia. Nat Genet 2004; 36:453-461.

96. Holtschke T, Lohler J, Kanno Y et al. Immunodeficiency and chronic myelogenous leukemia-like syndrome in mice with a targeted mutation of the ICSBP gene. Cell 1996; 87:307-317.

97. Passegue E, Jochum W, Schorpp-Kistner M et al. Chronic myeloid leukemia with increased granulocyte progenitors in mice lacking junB expression in the myeloid lineage. Cell 2001; 104:21-32.

98. Passegue E, Wagner EF, Weissman IL. JunB deficiency leads to a myeloproliferative disorder arising from hematopoietic stem cells. Cell 2004; 119:431-443.

99. Ishida D, Kometani K, Yang H et al. Myeloproliferative stem cell disorders by deregulated Rap1 activation in SPA-1-deficient mice. Cancer Cell 2003; 4:55-65.

100. Tamura T, Nagamura-Inoue T, Shmeltzer Z et al. ICSBP directs bipotential myeloid progenitor cells to differentiate into mature macrophages. Immunity 2000; 13:155-165.

101. Stork PJ. Does Rap1 deserve a bad Rap? Trends Biochem Sci 2003; 28:267-275.

102. Hao SX, Ren R. Expression of ICSBP is downregulated in Bcr-Abl-induced murine CML-like disease, and forced coexpression of ICSBP inhibits the Bcr-Abl-induced myeloproliferative disorder. Mol Cell Biol 2000; 20:1149-1161.

103. Schmidt M, Nagel S, Proba J et al. Lack of interferon consensus sequence binding protein (ICSBP) transcripts in human myeloid leukemias. Blood 1998; 91:22-29.

104. Bruchova H, Borovanova T, Klamova H et al. Gene expression profiling in chronic myeloid leukemia patients treated with hydroxyurea. Leuk Lymphoma 2002; 43:1289-1295.

105. Yang MY, Liu TC, Chang JG et al. JunB gene expression is inactivated by methylation in chronic myeloid leukemia. Blood 2003; 101:3205-3211.

106. Mizuchi D, Kurosu T, Kida A et al. BCR/ABL activates Rap1 and B-Raf to stimulate the MEK/Erk signaling pathway in hematopoietic cells. Biochem Biophys Res Commun 2005; 326:645-651.

107. McLaughlin J, Chianese E, Witte ON. Alternative forms of the BCR-ABL oncogene have quantitatively different potencies for stimulation of immature lymphoid cells. Mol Cell Biol 1989; 9:1866-1874.

108. Castellanos A, Pintado B, Weruaga E et al. A BCR-ABL(p190) fusion gene made by homologous recombination causes B- cell acute lymphoblastic leukemias in chimeric mice with independence of the endogenous bcr product. Blood 1997; 90:2168-2174.

109. Heisterkamp N, Jenster G, ten Hoeve J et al. Acute leukemia in bcr/abl transgenic mice. Nature 1990; 344:251-253.

110. Voncken JW, Griffiths S, Greaves MF et al. Restricted oncogenicity of BCR/ABL p190 in transgenic mice. Cancer Res 1992; 52:4534-4539.

111. Voncken JW, Kaartinen V, Pattengale PK et al. BCR/ABL P210 and P190 cause distinct leukemia in transgenic mice. Blood 1995; 86:4603-4611.

112. Huettner CS, Zhang P, Van Etten RA et al. Reveribility of acute B-cell leukemia induced by BCR-ABL1. Nature Genetics 2000; 24:57-60.

113. Gross AW, Ren R. Bcr-Abl has a greater intrinsic capacity than v-Abl to induce the neoplastic expansion of myeloid cells in vivo. Oncogene 2000; 19:6286-6296.

114. Million RP, Harakawa N, Roumiantsev S et al. A direct binding site for Grb2 contributes to transformation and leukemogenesis by the Tel-Abl (ETV6-Abl) tyrosine kinase. Mol Cell Biol 2004; 24:4685-4695.

115. Jousset C, Carron C, Boureux A et al. A domain of TEL conserved in a subset of ETS proteins defines a specific oligomerization interface essential to the mitogenic properties of the TEL-PDGFR beta oncoprotein. EMBO J 1997; 16:69-82.
116. Hannemann JR, McManus DM, Kabarowski JH et al. Haemopoietic transformation by the TEL/ABL oncogene. Br J Haematol 1998; 102:475-485.
117. Kim CA, Phillips ML, Kim W et al. Polymerization of the SAM domain of TEL in leukemogenesis and transcriptional repression. EMBO J 2001; 20:4173-4182.
118. Million RP, Aster J, Gilliland DG et al. The Tel-Abl (ETV6-Abl) tyrosine kinase, product of complex (9; 12) translocations in human leukemia, induces distinct myeloproliferative disease in mice. Blood 2002; 99:4568-4577.
119. Okuda K, Golub TR, Gilliland DG et al. p210BCR/ABL, p190BCR/ABL, and TEL/ABL activate similar signal transduction pathways in hematopoietic cell lines. Oncogene 1996; 13:1147-1152.
120. Wertheim JA, Miller JP, Xu L et al. The biology of chronic myelogenous leukemia: Mouse models and cell adhesion. Oncogene 2002; 21:8612-8628.
121. Ren R. Modeling the dosage effect of oncogenes in leukemogenesis. Curr Opin Hematol 2004; 11:25-34.
122. Okuda K, Oda A, Sato Y et al. Signal transduction and cellular functions of the TEL/ARG oncoprotein. Leukemia 2005.
123. Okuda K, Sato Y, Sonoda Y et al. The TEL/ARG leukemia oncogene promotes viability and hyperresponsiveness to hematopoietic growth factors. Int J Hematol 2004; 79:138-146.
124. Soekarman D, von Lindern M, Daenen S et al. The translocation (6; 9) (p23; q34) shows consistent rearrangement of two genes and defines a myeloproliferative disorder with specific clinical features. Blood 1992; 79:2990-2997.
125. von Lindern M, Fornerod M, van Baal S et al. The translocation (6; 9), associated with a specific subtype of acute myeloid leukemia, results in the fusion of two genes, dek and can, and the expression of a chimeric, leukemia-specific dek-can mRNA. Mol Cell Biol 1992; 12:1687-1697.
126. Ryan KJ, Wente SR. The nuclear pore complex: A protein machine bridging the nucleus and cytoplasm. Curr Opin Cell Biol 2000; 12:361-371.
127. Eudy JD, Ma-Edmonds M, Yao SF et al. Isolation of a novel human homologue of the gene coding for echinoderm microtubule-associated protein (EMAP) from the Usher syndrome type 1a locus at 14q32. Genomics 1997; 43:104-106.

CHAPTER 8

Abl Family Kinases in Mammalian Development

Eva Marie Y. Moresco

Abstract

Abl and Arg nonreceptor tyrosine kinases are widely expressed in mammals, where they contribute to the development of diverse organ and tissue systems. Deletion of *abl* or *arg* in mice reveals roles for the kinases in B and T lymphocyte development, neurulation, neuronal dendrite maintenance, synaptic plasticity, and osteoblast development. Double knockout *abl*$^{-/-}$*arg*$^{-/-}$ mice die as embryos, indicating that Abl and Arg also perform essential and overlapping functions during embryonic development. Abl and Arg contain domains for protein-protein interactions (SH3, SH2, proline-rich sequences, PY sequences), cytoskeleton binding (filamentous actin and microtubule binding domains), nuclear translocation (nuclear localization and export sequences), and DNA binding. Although a full understanding of their molecular interactions is still forthcoming, it is clear that Abl and Arg provide many cell types with all-in-one multifunctional signaling tools that serve as links between the cell surface and downstream pathways to both the cytoskeleton and nucleus.

Introduction

The mammalian Abl nonreceptor tyrosine kinase family consists of Abl and Arg (Abl-related gene, also called Abl2). *c-abl* was identified as the cellular (c for cellular) homologue of *v-abl* (v for viral), the transforming gene of the Abelson murine leukemia virus (AMulV), and encodes the nonreceptor tyrosine kinase Abl.[1-3] *c-abl* (hereafter *abl*) was later shown to be involved in human leukemias due to a chromosomal translocation resulting in fusion of N-terminal sequences of *Bcr* to *abl*, causing formation of the Philadelphia chromosome and *Bcr-abl* oncogenes.[2] Bcr-abl tyrosine kinases are constitutively active and transform lymphoid cells in culture and in mice.[4] Because of the involvement of *abl* in malignancy, a search for related genes was performed (actually using c-DNA probes to *v-abl*) and the *abl*-related gene (*arg*) was identified.[5,6]

Abl and Arg share extensive sequence and structural similarity (Fig. 1). Both have two predominant alternatively spliced forms, in mice designated Abl type I and IV and Arg type 1a and 1b, each containing one of two different first exons and resulting in distinct N-terminal peptide sequences.[5,7,8] The biological functions of each isoform remains unclear, although a few reports have documented differences in requirements for type I and type IV Abl in mediating differentiation or survival in response to lipopolysaccharide (LPS) treatment of lymphoid cell lines.[9,10] However, either Abl isoform can complement the lymphocyte defects (see below) of

*Eva Marie Y. Moresco c/o Anthony Koleske—Department of Molecular Biophysics and Biochemistry, Yale University School of Medicine, New Haven, Connecticut 06520, U.S.A. Email: anthony.koleske@yale.edu

Abl Family Kinases in Development and Disease, edited by Anthony Koleske.
©2006 Landes Bioscience and Springer Science+Business Media.

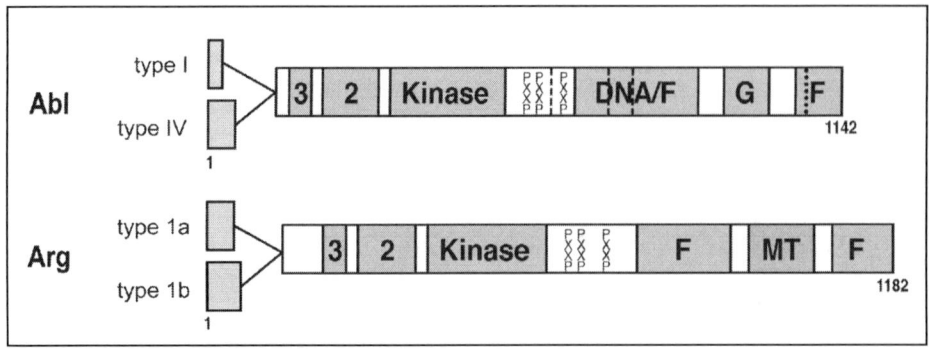

Figure 1. Domain structures of Abl and Arg. Abl and Arg each have two alternatively spliced first exons resulting in two expressed isoforms, Abl type I and IV, and Arg type 1a and 1b. Abl and Arg contain SH3 (3), SH2 (2) and kinase domains; these domains are 89%, 90% and 93% identical in Abl and Arg. In their C-terminal halves, Abl and Arg contain 3 conserved PXXP motifs, and binding domains for F-actin (F). Arg also contains a microtubule-binding domain (MT). Abl has a binding domain for globular actin (G), and a DNA-binding region (DNA) that overlaps with an internal F-actin binding domain. Nuclear localization signals (dashed lines) and nuclear export sequence (dotted line) are indicated in Abl. Amino acid numbers are listed at the N- and C-termini.

abl mutant mice when expressed as transgenes.[11] Following the N-terminal domain, Abl and Arg contain Src homology 3 (SH3), Src homology 2 (SH2) protein interaction domains and tyrosine kinase domains. They contain three conserved PXXP motifs C-terminal to the kinase domain that serve as potential binding sites for SH3 domain-containing proteins.[12] The near 90% identity in their N-terminal halves suggests that Abl and Arg have common interactors or substrates, and may share some cellular functions.

Abl and Arg also contain cytoskeleton-binding domains in their C-terminal halves, features unique among nonreceptor tyrosine kinases. These consist of filamentous (F)-actin binding domains in Abl and Arg[13-15] (M. Krishnaswami and A.J. Koleske, unpublished data), and a microtubule (MT) binding domain in Arg.[16] These domains localize Abl and Arg within the cell, and allow them to physically regulate actin and microtubule organization. Finally, Abl has three nuclear localization sequences (NLS)[17,18] and a nuclear export sequence (NES)[19] that are absent in Arg. These domains allow Abl to shuttle between the cytoplasm and the nucleus in response to various stimuli. A DNA binding domain identified in Abl[68] currently has un-known physiological function.

One look at the lengthy list of functional domains present in Abl and Arg suggests that their molecular interactors are numerous and their physiological functions complicated. Add to that the wide tissue expression pattern of Abl and Arg (see below), and it becomes clear that understanding their roles in development would not be easy. This has proven true. Since the generation of *abl* mutant mice nearly 15 years ago, how Abl kinases contribute to mammalian development has been and continues to be the subject of intense study. Mice lacking the second mammalian Abl family kinase member, Arg, were subsequently engineered and analysis of their phenotypes gives insight into the unique roles the kinases play during development. Examination of these mutant mice demonstrates that Abl and Arg participate in a wide and seemingly unrelated variety of developmental processes, and that Abl and Arg possess both shared and unique functions.

In this chapter, I review the developmental processes that require Abl and Arg function in mammals (mostly mice), which include lymphocyte development, neuronal development, and bone development. The diversity of these processes makes it impossible to thoroughly discuss every aspect of the signaling that regulates them. Therefore, in considering how Abl and Arg may participate in these molecular pathways, I attempted to focus on data obtained from experiments using primary tissues or cells, with the assumption that these results provide the best insight into the normal functions of the kinases. From this survey, it becomes clear that Abl and Arg are multifunctional tools with many features cells find useful. Cells of many types employ these functions for their specific cellular tasks. In this way, Abl and Arg are recruited to participate in signaling to transduce information from a wide variety of cell surface receptors to both the cytoskeleton and the nucleus.

Tissue Expression Patterns of Abl and Arg

Overlapping patterns of Abl and Arg expression are detected in many cell types and tissues, although the level of Abl expression may be quite different from that of Arg in a given tissue. In addition, their expression levels can vary substantially from tissue to tissue, and may change during the course of normal development. Abl is expressed in a wide range of mouse tissues, including the spleen, thymus and brain.[20-22] Abl can be isolated in purified synaptosomes from mouse brain.[23] Of adult human tissues, Abl is found in small mucous glands and ductules, gastric crypts, endocervix, ovarian follicles, breast myoepithelium, prostatic acini, renal tubules, transitional epithelium, skin adnexae, myeloid cells and osteoblasts.[24] In contrast to data obtained from mouse, no Abl is detected in human cerebral cortical neurons, lymph nodes or spleen.[24] Human fetal tissues show a similar expression pattern to the adult, with additional expression in squamous mucosa, sex cords, adrenal cortex and medulla, small intestinal and colonic epithelium, skeletal, cardiac and smooth muscle, capillary endothelial cells and endocardial intima. Strong expression is also observed in osteoblasts and their associated neovasculature in sites of endochondral ossification.[24]

Arg expression is highest in the brain, with significant levels also found in spleen, thymus and muscle tissue.[21,25] Arg is particularly concentrated in synapse-rich brain regions, and localizes to dendritic spines and to purified synaptosomes in adult mice.[21,26] Embryonic mice have high levels of both Abl and Arg in the developing neuroepithelium.[21] The variety of tissues expressing Abl and Arg suggests that many cellular processes and signaling pathways utilize these kinases to carry out their functions.

Deletion of *abl* and *arg* in Mice Reveals Essential Roles for the Kinases

Two *c-abl* mutant alleles were designed and utilized to engineer mice lacking Abl function. *abl²* (hereafter *abl⁻*) is a true null allele, and no Abl protein is detected in *abl²* homozygous (*abl⁻/⁻*) mice.[27] A second allele, designated *abl^m1*, eliminates the C-terminal one-third of Abl containing its DNA-binding and cytoskeleton-binding domains, but retains the N-terminal portion containing the SH3, SH2 and kinase domains, 3 proline-rich regions adjacent to the kinase domain, and NLS sequences.[28-30] Approximately 75% of *abl⁻/⁻* and *abl^m1/m1* mice are runted and die within the first two weeks after birth. The remaining 25% of *abl⁻/⁻* and *abl^m1/m1* mutants are highly susceptible to infections throughout life, such as pneumonia, gastroenteritis and infections of the nasal passages and ears.[27,28] The cause of increased susceptibility to infection in these mice remains to be demonstrated.

Homozygous mutant mice expressing a null allele of *arg* are strikingly healthy and robust compared to *abl* mutant mice.[21] Although *arg⁻/⁻* pups are runted at 3 weeks of age, by 6 weeks they are similar in weight to wild type littermates. *arg⁻/⁻* mice are born at expected Mendelian ratios, and survive to healthy adulthood.[21] Loss of Arg does not result in immune deficiency as seen in *abl* mutants.

Intercrosses of *abl* and *arg* mutants reveal overlapping roles for Abl and Arg during development.[21] While *abl⁺/⁻arg⁺/⁻* mice are born and survive in expected Mendelian ratios, only 60% of *abl⁺/⁻arg⁻/⁻* mice survive to adulthood. *abl⁻/⁻arg⁺/⁻* mice die embryonically at 15.5 days post-coitum (dpc) and display pericardial sac and peritoneal hemorrhaging. *abl⁻/⁻arg⁻/⁻* mice are most severely affected and die as embryos at 10.5 dpc. Internal bleeding in the pericardial sac is also observed in these embryos, and likely results in their early death.[21] However, the cellular and molecular defects that result in death of $abl^{-/-}arg^{-/-}$ and $abl^{-/-}arg^{+/-}$ embryos are yet unknown. Immunostaining revealed Abl expression in cardiac muscles of human fetuses,[24] suggesting that heart development or function may require Abl and Arg. Alternatively, deletion of Abl and Arg may slightly compromise the development of several systems, which when combined fail to support the life of the embryo.

Lymphocyte Development in $abl^{-/-}$ and $abl^{m1/m1}$ Mice

B and T Cell Development Require Abl

The bone marrow and thymus are the major sites for early development of the immune system's B and T lymphocytes, respectively. These organs provide both soluble and cell-anchored stromal signals that direct lymphocyte development, and as such, their unique environments are critical to the development of the immune system. Once they have completed their developmental programs there, naïve B and T lymphocytes leave the bone marrow and thymus, homing to peripheral lymphoid organs including the spleen and lymph nodes, where they complete their maturation and may become activated upon encountering antigen. As B and T cells work closely together to protect the body from invasion, both must properly develop and migrate in order for these cells to be available throughout the body to recognize and eliminate foreign antigens.

The increased susceptibility to infection of the $abl^{-/-}$ and $abl^{m1/m1}$ mice suggested that these mutants have defects in immune development or function. Indeed, similar (but not identical) phenotypes of reduced total B and T cell numbers are observed in $abl^{-/-}$ and $abl^{m1/m1}$ mutants.[27,28,31] Consistent with this finding, $abl^{-/-}$ and $abl^{m1/m1}$ mutant mice display a reduced size and cellularity of the thymus. Atrophy of the spleen is also observed, but is less severe than reductions in the thymus. In $abl^{m1/m1}$ mice, spleens exhibit an abnormal short, squatty shape with rounded edges, instead of the normal elongated knife-blade shape.[28]

Lymphocytes express specific combinations of immunoglobulin and other cell surface receptors as they progress through development. The combination of receptors serves as a means to identify populations of cells in specific developmental stages. For B cells, the developmental program progresses though pro-B, pre-B, and immature B stages before reaching the mature B stage capable of responding to antigens (Table 1). When examined by flow cytometry for stage-specific markers, adult $abl^{m1/m1}$ mice exhibit reductions in bone marrow-derived early B cell classes, including pro-B, pre-B and immature B cells (Table 1).[28,31] On average, total B cells from bone marrow are reduced to 67% of control B cell numbers, with more severe reductions in pre-B cells relative to pro-B cells.[31] In general, B cell depletion is less severe in spleen compared to bone marrow, and $abl^{m1/m1}$ mutants exhibit at worst a 19% reduction in splenic B cells. Interestingly, circulating mature B cells in peripheral blood are nearly normal in $abl^{m1/m1}$ mice, suggesting that the defect in early B cell development may be overcome in later stages of development.

The thymi of newborn $abl^{m1/m1}$ mice contain on average only one-third the number of cells found in those of control littermates, although the relative proportions of immature and mature T cells appear to be normal in these mutants. In contrast to the reduction in thymic T cells, the number of circulating mature T cells in peripheral blood is only slightly reduced, similar to the effect seen in B cells.[28] Splenic T cells are normal or elevated in $abl^{m1/m1}$ mice. Thus, T cell development is less severely affected by the abl^{m1} mutation than B cell development.

Like $abl^{m1/m1}$ mutants, $abl^{-/-}$ mice exhibit reductions in both B and T cell numbers.[27] Early B cell classes are reduced in $abl^{-/-}$ bone marrow, while mature B cells are present at normal levels relative to animal body weight. However, in contrast to $abl^{m1/m1}$ animals, reductions in total peripheral lymphocyte numbers to 73% of wild type levels are observed in 3 to 5 week old $abl^{-/-}$ mice. Deficiencies in both B and T cell populations contribute to this reduction; however, the specific populations of B or T cells affected has not been determined. Thus, very similar lymphocytic phenotypes are observed in two different abl mutants, providing solid evidence that Abl function contributes to the development of B and T lymphocytes.

Table 1. *Average percentage of B cells in abl$^{m1/m1}$ relative to control mice*

	Surface Markers	Bone Marrow	Spleen	Lymph Node	Peripheral Blood	Newborn Liver
Pro-B cells	B220$^+$CD43$^+$	16-128 (77)				
Pre-B cells	B220$^+$CD43$^-$IgM$^-$	1.2-210 (63)				100
B cells	B220$^+$CD43$^-$IgM$^+$	26-110 (67)	81-100	100	100	100
Immature	B220^{dull}IgM$^+$(IgD^{lo})	60	sl. reduction			100
Mature	B220^{bright}IgM$^+$(IgD^{hi})	sl. reduction				100

When a range is shown, the average is indicated in parentheses. Values are presented in or calculated from data in Schwartzberg et al, 1991; Hardin et al, 1995; Hardin et al, 1996.[27,28,31,55]

Does Arg contribute to lymphocyte development? Unlike *abl* mutants, no defects are observed in *arg*$^{-/-}$ lymphocyte numbers from bone marrow, thymus, spleen or peripheral blood.[21] A recent report raises the possibility that Arg cannot functionally compensate for Abl during early B cell development because its expression is low during pro-B and pre-B cell stages when defects are most severe in *abl* mutants, and only later increases in mature B cells.[32] However, because *abl*$^{-/-}$*arg*$^{-/-}$ mice die as embryos before significant lymphopoeisis take place, whether Arg and Abl serve redundant functions in lymphocyte development remains unknown.

How Does Abl Regulate Lymphocyte Development?

The stromal environments of the thymus and bone marrow provide soluble and membrane-anchored signals that direct lymphocyte development. Therefore, it is possible that mutation or deletion of *abl* results in an abnormal stromal environment that fails to support proper lymphocyte development. This hypothesis is supported by the observation that bone marrow transfer from *abl*$^{-/-}$ mice to irradiated syngeneic wild type recipients fails to recapitulate the lymphopenic phenotype of *abl*$^{-/-}$ donors.[27] In addition, long-term lymphoid bone marrow cultures containing normal numbers of lymphocytes and precursors, can be established from *abl*$^{m1/m1}$ bone marrow in vitro,[31] suggesting that the microenvironment in which lymphocytes develop in vivo is defective in *abl* mutants.

However, evidence also supports a cell autonomous role for Abl in regulating lymphocyte development. When *abl*$^{m1/m1}$ pro-B cells are cultured on a wild type stromal cell layer, fewer pro-B cells remain after six days in culture compared to wild type cells,[31] suggesting that intrinsic defects in *abl*$^{m1/m1}$ lymphocytes prevent their proliferation in a wild type environment. Furthermore, the phenotype of depleted pro-B and pre-B cells can be reconstituted in normal irradiated syngeneic hosts after bone marrow transfer from *abl*$^{m1/m1}$ mice,[28,31] further supporting the idea that defects in *abl*$^{m1/m1}$ lymphocytes are cell autonomous. The opposite experiment, to transplant wild type bone marrow into irradiated *abl* mutants, has not been performed. Together, these data are consistent with defects in both the stromal environment and the lymphocytes themselves contributing to lymphopenia in *abl*^{m1} mutants.

Interleukin-7, B Cell Receptor, and T Cell Receptor Signaling through Abl

Several studies suggest that Abl helps to mediate interleukin-7 (IL-7) signaling, providing an attractive mechanism to explain many of the observed lymphopoeisis defects in *abl* mutant mice (Fig. 2A). IL-7 is a stromal cell-derived cytokine critical for early bone marrow B cell development in mice. Similar to *abl* mutants, mice lacking IL-7 either through genetic

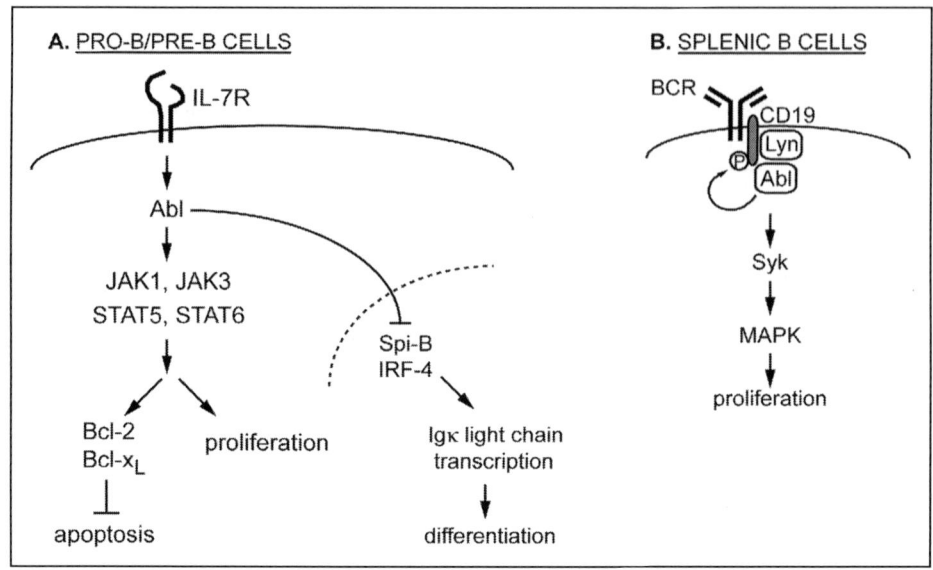

Figure 2. Hypothetical model for Abl regulation of B lymphocyte development. A) Abl may mediate signals from the IL-7 receptor (IL-7R). During the pro-B and pre-B cell stages, IL-7 signals activate JAK1, JAK3, STAT5 and STAT6, leading to proliferation and cell survival.[40] v-Abl can activate JAK1, JAK3, STAT5 and STAT6 in an IL-7-independent manner,[41] suggesting that in untransformed cells Abl may normally stimulate this pathway to promote proliferation and survival. In addition, v-Abl and IL-7 both positively regulate levels of the antiapoptotic proteins Bcl-2 and Bcl-x$_L$.[42] This finding suggests that increased pro-B and pre-B cell apoptosis observed in $abl^{m1/m1}$ mice may result from reduced levels of Bcl-2 and Bcl-x$_L$. Finally, v-Abl inhibits Spi-B and IRF-4, transcriptional activators of the Igk light chain locus, allowing differentiation and transition from pre-B to immature B cell stage.[66,67] Reduced numbers of pro-B and pre-B cells in $abl^{m1/m1}$ mutants may result from accelerated transition to more mature B cell stages[28] due to disinhibition of Spi-B and IRF-4. B) Abl may regulate BCR-mediated splenic B cell proliferation. Abl is found as a constitutive member of a lipid raft complex including CD19 (gray) and Lyn.[49] Abl phosphorylates Tyr490 on CD19,[49] potentially creating binding sites or phosphorylating other signaling molecules such as Syk and leading to MAP kinase activation and B cell proliferation.[50]

deletion or treatment with anti-IL-7 antibodies display defects in early lymphopoeisis (pro-B cell stage)[33,34] and consequent lymphopenia.[35] IL-7 stimulates proliferation of pro-B cells in vitro and in vivo.[36-38] This proliferative response to IL-7 is reduced in $abl^{m1/m1}$ bone marrow colony formation cultures relative to wild type cultures, although IL-7-induced proliferation of $abl^{m1/m1}$ cells in liquid culture is normal.[31] These observations suggest that IL-7-stimulated pro-B cell proliferation in bone marrow requires Abl function.

Further support for this hypothesis comes from studies of the activated v-Abl kinase expressed in AMulV-transformed cells. AMulV specifically transforms lymphoid precursors, including pro-B and pre-B cells, rendering their growth IL-7 independent.[39] Pre-B cells expressing v-Abl exhibit constitutively activated Janus protein kinase 1 (JAK1), JAK3, STAT5 and STAT6, critical elements of IL-7 signal transduction in pre-B cells.[40,41] Conversely, IL-7 treatment can reconstitute some v-Abl-mediated activities when v-Abl is acutely inactivated in transformed pre-B cells.[42] These results demonstrate that an activated Abl kinase deregulates and activates IL-7 signaling, leading to IL-7 independent, uncontrolled proliferation. Like expression of v-Abl, transgenic or exogenous IL-7 administration contributes to B lymphoid transformation.[43,44] It will be of interest to determine whether exogenous administration or transgenic IL-7 can overcome the lymphopenia of *abl* mutants.

Both $abl^{m1/m1}$ and IL-$7^{-/-}$ B cells also exhibit increased apoptosis, which may contribute to the observed reduction in pro-B and pre-B cells in mutant bone marrow.[45,46] Interestingly, whereas IL-7 withdrawal from wild type progenitor B-cell lines results in G1 arrest, it promotes apoptosis in $abl^{m1/m1}$ B-cell lines.[45,47] Furthermore, as IL-7 and v-Abl have been shown to positively regulate the levels of antiapoptotic proteins Bcl-2 and Bcl-x_L,[42] loss of Abl signaling may result in reduced levels of Bcl-2 and/or Bcl-x_L leading to increased apoptosis. Thus, Abl may mediate pro-survival signaling from the IL-7 receptor in bone marrow pro-B and pre-B cells.

Once B cells complete their development in the bone marrow, some will migrate to the spleen, where their survival and maintenance requires signals from the B cell receptor (BCR).[48] Interestingly, a recent study reports that $abl^{-/-}$ splenic B cells are deficient in their proliferative response to BCR stimulation, suggesting that depletion of B cells in the spleen of abl mutants may result from specific defects in the signaling pathway activated by BCR stimulation.[49] Abl associates with and phosphorylates the BCR coreceptor CD19, which serves to activate downstream components of BCR signaling by recruiting effector proteins such as Syk and Btk.[49,50] These data suggest that Abl and CD19 act in a similar signaling pathway that modulates BCR-mediated splenic B cell proliferation (Fig. 2B).

What about T cells? $abl^{m1/m1}$ and $abl^{-/-}$ mice display reductions in T cell numbers in the thymus and peripheral blood, respectively.[27,28] Treatment of chronic myelogenous leukemia (CML), a disease caused by expression of the activated Bcr-abl protein in myeloid cells, with the specific Abl and Arg kinase inhibitor STI-571 (imatinib mesylate or Gleevec), has recently prompted investigation of the effects of Abl inhibition on T cell proliferation. Even though T cells from CML patients do not express the STI-571 target Bcr-abl, their proliferation is inhibited in a dose-dependent manner by STI-571,[51,52] suggesting that Abl is required for the normal proliferation of T cells. In particular, T cell receptor (TCR)-mediated T cell proliferation is blocked by STI-571.[53,54] Importantly, recent work demonstrates that TCR stimulation activates Abl to phosphorylate Zap70 and the linker for activation of T cells (LAT), effectors of TCR signaling that lead to transcription and eventual production of the cytokine IL-2.[53] Thus, Abl can phosphorylate two downstream effectors of the TCR to stimulate a T cell proliferation pathway.

Immune Function in abl *Mutants*

The defects in B and T cell development may contribute to the susceptibility of $abl^{m1/m1}$ and $abl^{-/-}$ mice to infections. Surprisingly, although in vitro responses of $abl^{m1/m1}$ B and T cells to mitogens is reduced in several instances, the primary immune response of $abl^{m1/m1}$ mice to sheep red blood cells (SRBCs) is normal.[55] In this assay, mice are immunized with SRBCs and then tested to see if they produced antibodies to SRBC antigens. Thus, one unexplored possibility is that $abl^{m1/m1}$ mice can produce antibodies, but defects in lymphocyte migration prevent the full and effective immune response. A new study finds that Abl, the Abl interactor protein-1 (Abi-1), and WAVE2, a member of the Wiscott-Aldrich syndrome family, form a complex that promotes actin polymerization in Jurkat T cells.[56] As actin polymerization drives cell migration, deletion of Abl in T cells might compromise their ability to migrate to sites of infection.

Neuronal Development in $abl^{-/-}$ and $arg^{-/-}$ Mice

Neurulation and Dendrite Maintenance Require Abl and Arg

The entire central nervous system (CNS) develops from the neural tube, which forms when a sheet of ectodermal cells folds and fuses to become a fluid-filled tube in a process called neurulation. All of the specialized structures of the brain develop from this tube. Once this occurs, wiring the nervous system involves the formation of appropriate connections between neurons, a process that begins with the proliferation and migration of the neurons. From their final positions in the brain, neurons extend one axon and several dendrites, specialized compartments from which neurons send signals out and receive signals from other cells, respectively. Axons and dendrites make connections with each other at junctions called synapses, contact areas between a presynaptic axon from one neuron, and a postsynaptic dendrite from another neuron. Neurons are highly interconnected, and each neuron may have thousands of synapses. Because synapses are the basis of information transfer in the brain, the size and structure of axons and dendrites on which synapses form are important determinants of brain function.

Abl and Arg are highly expressed in neurons of the developing and adult mouse brain.[21,26,57] Abl has been localized to synapses [23,26] and to axonal growth cones.[57] Compared to other tissues, Arg is most highly expressed in the brain, where it is concentrated in synapses and neuronal processes.[21,26] For example, Arg is found in the synapse-rich molecular layers of the cerebellum and the hippocampus, but not in the cell body-containing granular layers of these structures.[21] *arg*[-/-] mice exhibit multiple behavioral abnormalities, including motor deficits and sensorineural deafness,[21] suggesting that Arg may regulate synaptic transmission. The expression levels of Abl and Arg are developmentally regulated. During embryogenesis, Abl is about 5-fold more abundant than Arg in whole brain extracts. Postnatally, Arg becomes about 8-fold more abundant than Abl at postnatal day 21 (P21), and 5-fold more abundant than Abl in the six-week-old brain (E.M.Y. Moresco and A.J. Koleske, unpublished data). Their differential expression levels may indicate different roles for Abl and Arg over the course of brain development.

The generation of *abl*[-/-]*arg*[-/-] double knockout mice revealed overlapping roles for Abl and Arg in embryonic central nervous system (CNS) development.[21] While single knockout *abl*[-/-] or *arg*[-/-] mice live to adulthood and display no gross brain abnormalities, *abl*[-/-]*arg*[-/-] embryos die before embryonic day 11 and exhibit several defects in neurulation. Closure of the neural tube is delayed in *abl*[-/-]*arg*[-/-] embryos. Then, as the neural tube of these mice proceeds towards closure, the neuroepithelium buckles into the lumen of the neural tube. Abl and Arg colocalize with each other and with actin filaments at the apical surface of the developing neuroepithelium, and this actin network is disrupted in *abl*[-/-]*arg*[-/-] embryos.[21] It remains unclear whether the actin cytoskeletal defects cause the failure in neural tube closure, or whether they are the result of other unknown defects in these embryos. Accumulating evidence suggests that Abl and Arg regulate actin cytoskeletal organization,[58,59] supporting the hypothesis that actin defects in the neuroepithelium cause the failure in neurulation.

Our recent studies analyzed the structure of cortical neurons in *abl*[-/-], *arg*[-/-] and conditional brain-specific (bs)-*abl*[-/-]*arg*[-/-] double knockout mice, and found that dendrite arbor maintenance is deficient in these mutants.[60] Cortical dendrites initially develop normally in *arg*[-/-] and *bs-abl*[-/-]*arg*[-/-] mice, but by early adulthood mutant dendrites are reduced in size and complexity relative to wild type dendrites. Interestingly, adult *bs-abl*[-/-]*arg*[-/-] dendrites are more severely affected than either single mutant, suggesting that Abl and Arg have overlapping functions in dendrite maintenance. Using primary cortical neurons from mutant mice, we demonstrated that Arg is required for neurite branching stimulated by integrin-dependent cell adhesion to the ligands Semaphorin7A (Sema7A) or laminin-1. Together, the data suggest that Abl and Arg regulate dendrite branch maintenance in response to adhesive cues.

The downstream effectors of Abl and Arg signaling in dendrite maintenance are currently unknown. However, one good candidate is the 190kD GTPase activating protein for RhoA (p190RhoGAP), a neuronal substrate of Arg during postnatal periods.[61] p190RhoGAP inhibits the RhoA GTPase, which has been shown to negatively regulate dendrite branching in other systems.[62] In the future, examining the dendrites of *sema7a* and *p190rhogap* mutants, and crossing these mutations into *abl*[-/-] or *arg*[-/-] background, should provide insight into whether these proteins function in similar pathways to regulate dendrite morphology.

Regulation of Synapse Structure and Function by Abl and Arg

Emerging evidence suggests that Abl family kinases regulate synaptic structure and function. Abl family kinases regulate the assembly of postsynaptic components at the neuromuscular junction (NMJ).[63] Abl and Arg are required for the agrin-induced clustering of acetylcholine receptors (AchRs) on the postsynaptic membrane, where they form a complex with and phosphorylate the muscle-specific receptor tyrosine kinase (MuSK). Thus, Abl and Arg regulate neurotransmitter receptor distribution in the postsynaptic compartment of the NMJ.

In the CNS, Abl and Arg also localize to both presynaptic terminals and dendritic spines, the postsynaptic compartments of excitatory synapses.[26] Electrophysiological studies in the mouse hippocampus reveal that Abl and Arg modulate the efficiency of neurotransmitter release from the presynaptic terminal.[26] Paired-pulse facilitation (PPF), a transient form of presynaptic plasticity, is reduced in *abl*$^{-/-}$ and *arg*$^{-/-}$ hippocampal slices, and in wild type slices treated acutely with STI-571. Interestingly, treatment of *abl*$^{-/-}$ or *arg*$^{-/-}$ slices with STI-571 did not further reduce PPF, indicating that Abl and Arg have unique roles in the synapse and both are required for PPF. These data suggest that Abl and Arg kinase activity support optimal neurotransmitter release from the presynaptic terminal.

The mechanisms by which Abl and Arg regulate synaptic function are currently unknown. It is likely that they mediate interactions between cell surface receptors and the cytoskeleton in synapses, but it is unclear with which cell surface receptors they interface. Cell adhesion receptors are likely candidates because of their established roles in the formation, maintenance and remodeling of synaptic contacts. Investigating the synaptic partners of Abl and Arg will be an important area for future research.

Bone Development in *abl*$^{-/-}$ Mice

Consistent with the high expression level of Abl in osteoblasts of human fetuses,[24] *abl*$^{-/-}$ mice have osteoporosis, displaying thinner bone volume and bone mineral content than wild type mice.[64] Thinning of bone volume may be due to increased bone absorption by osteoclasts or reduced bone deposition by osteoblasts. Therefore, proper development and function of osteoclasts and osteoblasts is critical for bone development and maintenance. Osteoclastogenesis and osteoclast function are normal in *abl*$^{-/-}$ mice, but by several measures, osteoblastogenesis is delayed or defective during early maturation.[64] This leads to a dramatically reduced rate of bone mineral deposition in *abl*$^{-/-}$ mice.

Bone development is probably the least understood of all *abl* mutant phenotypes. There have been no further studies on Abl regulation of osteoblast development, or any reports of whether Arg may contribute to this process. Li et al[64] speculate that Abl may participate in extracellular matrix (ECM)-integrin signaling in osteoblasts, a critical pathway in osteoblast differentiation. This hypothesis is supported by the finding that Abl kinase activity increases upon fibroblast adhesion to the ECM protein fibronectin,[65] and by our work demonstrating that Abl mediates integrin signaling in developing neurons.[60]

Conclusion

The study of *abl* and *arg* mutant mice has certainly advanced our understanding of the normal functions of Abl and Arg. We now know that Abl and Arg participate in a wide variety of developmental processes. However, we have yet to fully understand the molecular mechanisms the kinases use to regulate these processes. In addition, the molecular basis of overlapping versus distinct functions of Abl and Arg in specific processes is unknown. Many mutant phenotypes occur with variable penetrance and severity. This has been noted for several aspects of lymphocyte development,[27,28,31,55] as well as bone development,[64] but its cause also remains unknown. Future work will undoubtedly shed light on these questions.

There seems to be no rhyme or reason to the group of tissue systems in which Abl and Arg function during development, except that in each developmental event, Abl and Arg have been utilized by a specialized cell type for a specific signaling need. For Abl and Arg, these range from pathways promoting cell proliferation and survival to cytoskeletal regulation. Clearly, with the versatility of a Swiss Army knife, Abl and Arg are perfectly suited for this multi-tasking.

References

1. Goff SP, Gilboa E, Witte ON et al. Structure of the Abelson murine leukemia virus genome and the homologous cellular gene: studies with cloned viral DNA. Cell 1980; 22(3):777-785.
2. Shtivelman E, Lifshitz B, Gale RP et al. Fused transcript of abl and bcr genes in chronic myelogenous leukaemia. Nature 1985; 315(6020):550-554.
3. Abelson HT, Rabstein LS. Lymphosarcoma: virus-induced thymic-independent disease in mice. Cancer Res 1970; 30(8):2213-2222.
4. Van Etten RA. Studying the pathogenesis of BCR-ABL+ leukemia in mice. Oncogene 2002; 21(56):8643-8651.
5. Kruh GD, Perego R, Miki T et al. The complete coding sequence of arg defines the Abelson subfamily of cytoplasmic tyrosine kinases. Proc Natl Acad Sci USA 1990; 87(15):5802-5806.
6. Kruh GD, King CR, Kraus MH et al. A novel human gene closely related to the abl proto-oncogene. Science 1986; 234(4783):1545-1548.
7. Shtivelman E, Lifshitz B, Gale RP et al. Alternative splicing of RNAs transcribed from the human abl gene and from the bcr-abl fused gene. Cell 1986; 47(2):277-284.
8. Ben-Neriah Y, Bernards A, Paskind M et al. Alternative 5' exons in c-abl mRNA. Cell 1986; 44(4):577-586.
9. Daniel R, Chung SW, Eisenstein TK et al. Specific association of Type I c-Abl with Ran GTPase in lipopolysaccharide-mediated differentiation. Oncogene 2001; 20(21):2618-2625.
10. Daniel R, Wong PM, Chung SW. Isoform-specific functions of c-abl: type I is necessary for differentiation, and type IV is inhibitory to apoptosis. Cell Growth Differ 1996; 7(9):1141-1148.
11. Hardin JD, Boast S, Mendelsohn M et al. Transgenes encoding both type I and type IV c-abl proteins rescue the lethality of c-abl mutant mice. Oncogene 1996; 12(12):2669-2677.
12. Pendergast AM. The Abl family kinases: mechanisms of regulation and signaling. Adv Cancer Res 2002; 85:51-100.
13. McWhirter JR, Wang JY. An actin-binding function contributes to transformation by the Bcr-Abl oncoprotein of Philadelphia chromosome-positive human leukemias. EMBO J 1993; 12(4):1533-1546.
14. Van Etten RA, Jackson PK, Baltimore D et al. The COOH terminus of the c-Abl tyrosine kinase contains distinct F- and G-actin binding domains with bundling activity. J Cell Biol 1994; 124(3):325-340.
15. Wang Y, Miller AL, Mooseker MS et al. The Abl-related gene (Arg) nonreceptor tyrosine kinase uses two F-actin-binding domains to bundle F-actin. Proc Natl Acad Sci USA 2001; 98(26):14865-14870.
16. Miller AL, Wang Y, Mooseker MS et al. The Abl-related gene (Arg) requires its F-actin-microtubule cross-linking activity to regulate lamellipodial dynamics during fibroblast adhesion. J Cell Biol 2004; 165(3):407-419.
17. Wen ST, Jackson PK, Van Etten RA. The cytostatic function of c-Abl is controlled by multiple nuclear localization signals and requires the p53 and Rb tumor suppressor gene products. EMBO J 1996; 15(7):1583-1595.
18. Van Etten RA, Jackson P, Baltimore D. The mouse type IV c-abl gene product is a nuclear protein, and activation of transforming ability is associated with cytoplasmic localization. Cell 1989; 58(4):669-678.
19. Taagepera S, McDonald D, Loeb JE et al. Nuclear-cytoplasmic shuttling of C-ABL tyrosine kinase. Proc Natl Acad Sci USA 1998; 95(13):7457-7462.
20. Muller R, Slamon DJ, Tremblay JM et al. Differential expression of cellular oncogenes during pre- and postnatal development of the mouse. Nature 1982; 299(5884):640-644.
21. Koleske AJ, Gifford AM, Scott ML et al. Essential roles for the Abl and Arg tyrosine kinases in neurulation. Neuron 1998; 21(6):1259-1272.
22. Renshaw MW, Capozza MA, Wang JY. Differential expression of type-specific c-abl mRNAs in mouse tissues and cell lines. Mol Cell Biol 1988; 8(10):4547-4551.
23. Courtney KD, Grove M, Vandongen H et al. Localization and phosphorylation of Abl-interactor proteins, Abi-1 and Abi-2, in the developing nervous system. Mol Cell Neurosci 2000; 16(3):244-257.
24. O'Neill AJ, Cotter TG, Russell JM et al. Abl expression in human fetal and adult tissues, tumours, and tumour microvessels. J Pathol 1997; 183(3):325-329.
25. Perego R, Ron D, Kruh GD. Arg encodes a widely expressed 145 kDa protein-tyrosine kinase. Oncogene 1991; 6(10):1899-1902.
26. Moresco EM, Scheetz AJ, Bornmann WG et al. Abl family nonreceptor tyrosine kinases modulate short-term synaptic plasticity. J Neurophysiol 2003; 89(3):1678-1687.

27. Tybulewicz VL, Crawford CE, Jackson PK et al. Neonatal lethality and lymphopenia in mice with a homozygous disruption of the c-abl proto-oncogene. Cell 1991; 65(7):1153-1163.

28. Schwartzberg PL, Stall AM, Hardin JD et al. Mice homozygous for the ablm1 mutation show poor viability and depletion of selected B and T cell populations. Cell 1991; 65(7):1165-1175.

29. Schwartzberg PL, Goff SP, Robertson EJ. Germ-line transmission of a c-abl mutation produced by targeted gene disruption in ES cells. Science 1989; 246(4931):799-803.

30. Schwartzberg PL, Robertson EJ, Goff SP. Targeted gene disruption of the endogenous c-abl locus by homologous recombination with DNA encoding a selectable fusion protein. Proc Natl Acad Sci USA 1990; 87(8):3210-3214.

31. Hardin JD, Boast S, Schwartzberg PL et al. Bone marrow B lymphocyte development in c-abl-deficient mice. Cell Immunol 1995; 165(1):44-54.

32. Bianchi C, Muradore I, Corizzato M et al. The expression of the nonreceptor tyrosine kinases Arg and c-abl is differently modulated in B lymphoid cells at different stages of differentiation. FEBS Lett 2002; 527(1-3):216-222.

33. Peschon JJ, Morrissey PJ, Grabstein KH et al. Early lymphocyte expansion is severely impaired in interleukin 7 receptor-deficient mice. J Exp Med 1994; 180(5):1955-1960.

34. Grabstein KH, Waldschmidt TJ, Finkelman FD et al. Inhibition of murine B and T lymphopoiesis in vivo by an anti-interleukin 7 monoclonal antibody. J Exp Med 1993; 178(1):257-264.

35. von Freeden-Jeffry U, Vieira P, Lucian LA et al. Lymphopenia in interleukin (IL)-7 gene-deleted mice identifies IL-7 as a nonredundant cytokine. J Exp Med 1995; 181(4):1519-1526.

36. Valenzona HO, Dhanoa S, Finkelman FD et al. Exogenous interleukin 7 as a proliferative stimulant of early precursor B cells in mouse bone marrow: efficacy of IL-7 injection, IL-7 infusion and IL-7-anti-IL-7 antibody complexes. Cytokine 1998; 10(6):404-412.

37. Namen AE, Lupton S, Hjerrild K et al. Stimulation of B-cell progenitors by cloned murine interleukin-7. Nature 1988; 333(6173):571-573.

38. Morrissey PJ, Conlon P, Charrier K et al. Administration of IL-7 to normal mice stimulates B-lymphopoiesis and peripheral lymphadenopathy. J Immunol 1991; 147(2):561-568.

39. Rosenberg N. Abl-mediated transformation, immunoglobulin gene rearrangements and arrest of B lymphocyte differentiation. Semin Cancer Biol 1994; 5(2):95-102.

40. Darnell JE Jr, Kerr IM, Stark GR. Jak-STAT pathways and transcriptional activation in response to IFNs and other extracellular signaling proteins. Science 1994; 264(5164):1415-1421.

41. Danial NN, Pernis A, Rothman PB. Jak-STAT signaling induced by the v-abl oncogene. Science 1995; 269(5232):1875-1877.

42. Banerjee A, Rothman P. IL-7 reconstitutes multiple aspects of v-Abl-mediated signaling. J Immunol 1998; 161(9):4611-4617.

43. Fisher AG, Burdet C, Bunce C et al. Lymphoproliferative disorders in IL-7 transgenic mice: expansion of immature B cells which retain macrophage potential. Int Immunol 1995; 7(3):415-423.

44. Rich BE, Campos-Torres J, Tepper RI et al. Cutaneous lymphoproliferation and lymphomas in interleukin 7 transgenic mice. J Exp Med 1993; 177(2):305-316.

45. Dorsch M, Goff SP. Increased sensitivity to apoptotic stimuli in c-abl-deficient progenitor B-cell lines. Proc Natl Acad Sci USA 1996; 93(23):13131-13136.

46. Lu L, Osmond DG. Apoptosis and its modulation during B lymphopoiesis in mouse bone marrow. Immunol Rev 2000; 175:158-174.

47. Griffiths SD, Goodhead DT, Marsden SJ et al. Interleukin 7-dependent B lymphocyte precursor cells are ultrasensitive to apoptosis. J Exp Med 1994; 179(6):1789-1797.

48. Lam KP, Kuhn R, Rajewsky K. In vivo ablation of surface immunoglobulin on mature B cells by inducible gene targeting results in rapid cell death. Cell 1997; 90(6):1073-1083.

49. Zipfel PA, Grove M, Blackburn K et al. The c-Abl tyrosine kinase is regulated downstream of the B cell antigen receptor and interacts with CD19. J Immunol 2000; 165(12):6872-6879.

50. Reth M, Wienands J. Initiation and processing of signals from the B cell antigen receptor. Annu Rev Immunol 1997; 15:453-479.

51. Cwynarski K, Laylor R, Macchiarulo E et al. Imatinib inhibits the activation and proliferation of normal T lymphocytes in vitro. Leukemia 2004; 18(8):1332-1339.

52. Dietz AB, Souan L, Knutson GJ et al. Imatinib mesylate inhibits T-cell proliferation in vitro and delayed-type hypersensitivity in vivo. Blood 2004; 104(4):1094-1099.

53. Zipfel PA, Zhang W, Quiroz M et al. Requirement for Abl kinases in T cell receptor signaling. Curr Biol 2004; 14(14):1222-1231.

54. Seggewiss R, Lore K, Greiner E et al. Imatinib inhibits T-cell receptor mediated T-cell proliferation and activation in a dose-dependent manner. Blood 2005; 105(6):2473-9.

55. Hardin JD, Boast S, Schwartzberg PL et al. Abnormal peripheral lymphocyte function in c-abl mutant mice. Cell Immunol 1996; 172(1):100-107.

56. Leng Y, Zhang J, Badour K et al. Abelson-interactor-1 promotes WAVE2 membrane translocation and Abelson-mediated tyrosine phosphorylation required for WAVE2 activation. Proc Natl Acad Sci USA 2005; 102(4):1098-1103.
57. Zukerberg LR, Patrick GN, Nikolic M et al. Cables links Cdk5 and c-Abl and facilitates Cdk5 tyrosine phosphorylation, kinase upregulation, and neurite outgrowth. Neuron 2000; 26(3):633-646.
58. Woodring PJ, Hunter T, Wang JY. Regulation of F-actin-dependent processes by the Abl family of tyrosine kinases. J Cell Sci 2003; 116(Pt 13):2613-2626.
59. Hernandez SE, Krishnaswami M, Miller AL et al. How do Abl family kinases regulate cell shape and movement? Trends Cell Biol 2004; 14(1):36-44.
60. Moresco EMY, Donaldson S, Williamson A et al. Integrin-mediated dendrite maintenance requires Abl family kinases. J Neurosci. 2005; 25(26):6105-6118.
61. Hernandez SE, Settleman J, Koleske AJ. Adhesion-dependent regulation of p190RhoGAP in the developing brain by the Abl-related gene tyrosine kinase. Curr Biol 2004; 14(8):691-696.
62. Ruchhoeft ML, Ohnuma S, McNeill L et al. The neuronal architecture of Xenopus retinal ganglion cells is sculpted by rho-family GTPases in vivo. J Neurosci 1999; 19(19):8454-8463.
63. Finn AJ, Feng G, Pendergast AM. Postsynaptic requirement for Abl kinases in assembly of the neuromuscular junction. Nat Neurosci 2003; 6(7):717-723.
64. Li B, Boast S, de los Santos K et al. Mice deficient in Abl are osteoporotic and have defects in osteoblast maturation. Nat Genet 2000; 24(3):304-308.
65. Lewis JM, Schwartz MA. Integrins regulate the association and phosphorylation of paxillin by c-Abl. J Biol Chem 1998; 273(23):14225-14230.
66. Kurosaki T. Checks and balances on developing B cells. Nat Immunol 2003; 4(1):13-15.
67. Muljo SA, Schlissel MS. A small molecule Abl kinase inhibitor induces differentiation of Abelson virus-transformed pre-B cell lines. Nat Immunol 2003; 4(1):31-37.
68. Miao YJ, Wang JY. Binding of A/T-rich DNA by three high mobility group-like domains in c-Abl tyrosine kinase. J Biol Chem 1996; 271(37):22823-22830.

CHAPTER 9

Abelson Family Protein Tyrosine Kinases and the Formation of Neuronal Connectivity

Cheryl L. Thompson and David Van Vactor*

Introduction

T he nervous system is an organ of immense complexity. Neural function and the integration of neural input depend upon the formation of an intricate network of synaptic connections. Building this neural architecture during development involves several aspects of neuronal morphogenesis, from neuronal polarization and the extension of neuronal processes, to the pathfinding of axons across long distances to appropriate target cells and the elaboration of dendritic arbors, resulting in the assembly and maintenance of synapses between each neuron and its targets. Each neuronal subclass relies upon multiple extracellular cues to direct its pathfinding and synaptogenesis to the correct locations within the developing embryo. While much progress has been made in the identification of the extracellular factors and corresponding cell surface receptors that control these aspects of neuronal differentiation, much less is known about the intracellular molecules and signaling pathways that control the process of morphogenesis.[1]

As in nonneuronal cell biology, protein tyrosine phosphorylation plays an important role in controlling neuronal morphogenesis downstream of extracellular factors. Recent evidence from multiple neural systems indicates that Abelson (Abl) family protein tyrosine kinases (PTKs) are important for axon guidance, dendritic morphogenesis, and synaptogenesis. This work and parallel studies in nonneuronal cells have implicated Abl as a molecular link that controls the cytoskeletal biology of the neuron.

In particular, accumulated data suggest that Abl serves as a key intracellular control point upstream of several different classes of cytoskeletal effector proteins and downstream of several classes of cell surface receptors. In this way, Abl is likely to be a signaling node that simultaneously coordinates the multiple aspects of neuronal cell biology necessary to accomplish complex morphogenetic tasks from the initial outgrowth of axons and dendrites toward appropriate target cells, to the establishment of synaptic connections that will support neuronal communication throughout the life of the organism. In the following sections, we will review many of the studies that first identified Abl as a player in neural development, along with a few examples of neural signaling pathways in which Abl's role has been best explored, emphasizing the contribution of *Drosophila melanogaster* as the genetic model system.

*Corresponding Author: David Van Vactor—Department of Cell Biology, Program in Neuroscience, and Harvard Center for Neurodegeneration and Repair, Harvard Medical School, 240 Longwood Avenue, Boston, Massachussetts 02115, U.S.A.
Email: david_vanvactor@hms.harvard.edu

Abl Family Kinases in Development and Disease, edited by Anthony Koleske.
©2006 Landes Bioscience and Springer Science+Business Media.

Abelson Tyrosine Kinases and Axon Guidance

Analysis of Abl Function in the Drosophila Nervous System

The genetic analysis of Abl-family kinase function in mammals revealed the existence of two related vertebrate genes, c-Abl and the Abl-related gene (Arg), which appear to act with some degree of functional redundancy during neural development.[2] Indeed, *Abl/Arg* double null mice present gross defects in brain development and embryonic lethality, presenting an obstacle for functional analysis.[2] The molecular cloning of a single Abl homolog from the *Drosophila* genome with approximately equal sequence identity to mammalian c-Abl and Arg demonstrated that the Abelson tyrosine kinase family was highly conserved and identified *Drosophila* as a potential model system for the dissection of Abl signaling in vivo.[3,4] Mutations in *Drosophila Abl* display pleiotropic phenotypes during development, suggesting that Abl kinase is involved in a variety of signaling pathways.[5] Abl is an essential gene in *Drosophila*; yet, zygotic *Abl* mutants display a late lethal phase (during pupariation), raising the question of how the early embryo and larva can function in the absence of Abl activity? Interestingly, subsequent analysis of mutants lacking both maternal and zygotic Abl gene product revealed that maternally supplied Abl provides the embryo with enough protein to avoid major developmental defects.[6]

Despite abundant expression of Abl during CNS development, the initial analysis of *Abl* zygotic mutants failed to identify obvious defects in the embryonic nervous system, leading to a hypothesis that functional redundancy with other tyrosine kinases might mask an early requirement for Abl.[7] This hypothesis challenged the notion that Abl might have a specific function that could be distinguished from other intracellular tyrosine kinases in the Src family. However, a more recent analysis of *Abl* null mutations using the cell surface marker Fasciclin II, which is specific to a subset of axons, revealed consistent defects in the pathfinding of embryonic motor growth cones to their muscle targets.[8]

Abl Function during Motor Axon Pathfinding

The first analysis of axon pathway defects in *Abl* zygotic null mutant embryos was performed in the neuromuscular system.[8] In wild type embryos, motor axons emerge from each segment of the *Drosophila* CNS in two major bundles: the segmental (SN) and intersegmental (ISN) nerves.[9] Once in the periphery, the intersegmantal nerve defasciculates or separates into three axon bundles; the ISN axons target the dorsal muscles, the ISNb axons target the ventral longitudinal muscles, and ISNd axons innervate the ventral oblique body wall muscles (Fig. 1A). The defasciculation of motor axons is closely associated with specific locations or "choice points" where each nerve branch extends to contact its target domain.[10] Once the leading edge of the motor axon or growth cone has entered the correct target domain, they extend towards the distal muscles and begin a process of filopodial exploration before each neuron selects its appropriate synaptic partner by the retraction of exuberant processes.[11] By embryonic stage 17, nine ISNb axons innervate specific muscles within the ventral region (6, 7, 12-14, 28 and 30) as the process of synaptogenesis begins (Fig. 1A).

In *Abl* zygotic null mutant embryos, ISNb axons enter the ventral longitudinal target domain, however, they frequently arrest before reaching the most distal muscles (Fig. 1B).[8] This ISNb "stop short" phenotype is also observed in mutants lacking several cytoskeletal signaling and effector proteins, including the cytoskeleton-binding proteins Stop short/Kakapo[10,12] and Profilin,[8,10] as well as the guanine-nucleotide exchange factor Trio.[13] The similarity of this growth cone arrest phenotype to that observed in mutants lacking cytoskeletal proteins raised the possibility that loss of Abl results in a failure of axon extension as opposed to a defect in growth cone directional specificity.[8]

One great advantage of *Drosophila* as a model system for the study of neural development is its accessibility to comprehensive genetic screens. While accumulation of mutations with identical phenotypes has always been a powerful tool to assemble genetic pathways, another approach to

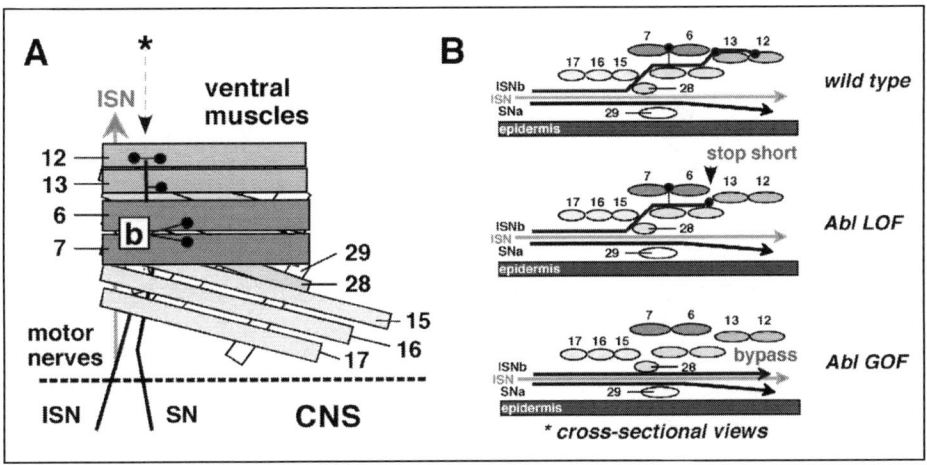

Figure 1. Abl is required for ventral motor axon guidance in Drosophila. A) Drosophila motor axons exit the CNS from two roots, the intersegmental and segmental nerves (ISN and SN). ISN branch b defasciculates from the ISN and enters the ventral muscle domain (muscles 7, 6, 13, 12, 14, 30 and 28). B) Cross-sectional views of wild type, Abl loss-of-function (LOF), and Abl gain-of-function (GOF) show the stop short and bypass phenotypes of ISNb motor axons relative to the muscle target cell positions.

pathway dissection is to search for dominant, dose-sensitive genetic interactions. Modifier screens are difficult to perform in vertebrate genetic models like zebrafish and mouse, yet they have been an effective approach for pathway analyses in *Drosophila*. Genetic modifier screens have been particularly useful in the case of Abl, since the adult phenotypes are highly pleiotropic[5] and the embryonic phenotypes are difficult to assay in a high-throughput fashion.[8] The first screen for Abl interactors took advantage of the late lethal phase of *Drosophila Abl* mutants in a loss-of-function approach (Fig. 2A). By mutagenizing an *Abl* mutant background and looking for haploinsufficiency at loci that would shift the lethal phase of an *Abl* homozygote, it was possible to identify genetic enhancers that made *Abl* mutants die early, or a suppressor that allowed *Abl* mutants to survive to adulthood.[14] This important pioneering screen for modifiers of *Abl* identified several candidate genes (Fig. 2B), including, the high-copy suppressor and substrate of Abl, Disabled,[14,15] in addition to the Abl-antagonist and substrate protein Enabled.[16,17] Other genes in this original collection of genetic modifiers included Prospero,[15] failed axonal connections (FAX),[18] the guanine nucleotide exchange factor Trio,[19] and the cell surface receptor Neurotactin (Nrt).[20]

Using double mutant analysis to examine dose-dependent genetic interactions, both *Drosophila* Profilin (*chickadee* or *chic*) and Trio demonstrated marked enhancement of *Abl* mutant phenotypes.[8,19] For example, many zygotic *chic;Abl* double mutants fail to extend motor axons to a greater extent than *Abl* single mutants,[8] and loss of maternally supplied Abl has similar exacerbation on the *Abl* mutant motor axon phenotype.[6] Further, the *Trio* mutant phenotype of discontinuous and thinning axons in the CNS was enhanced with the reduction of Abl activity.[19] Although the *Trio;Abl* double mutant study focused on the axons of the CNS, the axons in both instances still fail to extend normally when Abl dosage is reduced. Based on these and other allied data, it is likely that the Abl signaling pathway is part of core machinery essential to growth cone motility. Perturbation of Abl activity in vertebrate neuron cultures also disrupts normal axon outgrowth, suggesting conservation across species.[21] Additional clues as to the role of Abl signaling in motor axon guidance have come from links between Abl and cell surface receptors. Tests for dose-sensitive genetic interaction have identified a number of receptors and cell adhesion molecules that interact with Abl,[20,22-24] Notch,[25,26] the Robo

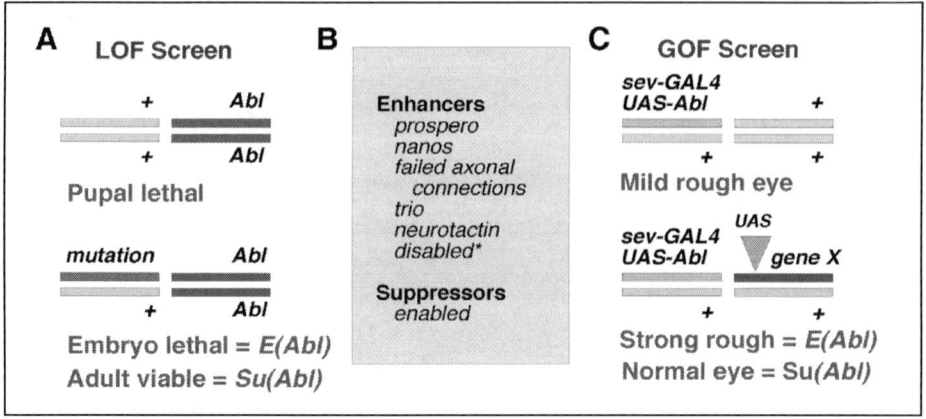

Figure 2. Genetic screens to dissect the Drosophila Abl pathway. A) The first loss-of-function (LOF) "F2" genetic interaction screen for Abl signaling partners in *Drosophila* was designed to isolate dominant mutations (see mutation chromosome) that would alter the lethal phase of Abl homozygous animals (Abl allele). Enhancer and suppressor loci on all chromosomes could be recovered with this scheme (see Gertler et al, 1989, 1990). B) A list of interacting loci recovered from the original LOF screens (see text for references). *Note that disabled acts as a high-copy suppressor of Abl phenotypes, but that the disabled alleles originally described are now thought to represent mutations in neurotactin.[20] C) An alternative and more efficient "F1" dominant interaction screen was made possible with GAL4 technology to identify modifiers of an Abl gain-of-function (GOF) retinal phenotype.[28,61] Over-expression in the retina using either sevenless (sev) or GMR promotor fusions with GAL4 (a recombinant chromosome with sev-GAL4 and UAS-Abl insertions) results in a rough eye phenotype that can be modified by the introduction of a UAS target site upstream of genes (e.g., Orbit/MAST) that interact with Abl (inverted triangle represents insertion site adjacent to interacting loci).

receptor family (discussed in the section on midline axon guidance),[27-29] and the receptor protein tyrosine phosphatase LAR.[30] The possibility of a reciprocal relationship between a tyrosine kinase and tyrosine phosphatase driving motor axon guidance lead to further examination of LAR.[30]

Drosophila Abl and LAR: Tyrosine Phosphorylation Plays a Key Role in Motor Axon Guidance

Drosophila LAR (Dlar) is a well-conserved member of a receptor protein tyrosine phophatase (RPTP) family that includes three mammalian orthologs, LAR, PTPδ and PTPσ.[31] Loss of *Drosophila* LAR results in an ISNb "bypass" phenotype where ISNb motor axons fail to turn into the ventral domain, and instead follow ISN axons towards dorsal targets.[32] The penetrance of this phenotype is incomplete,[32,33] presumably due to functional redundancy with the other *Drosophila* receptor protein tyrosine receptor DPTP69D.[34]

Since tyrosine phosphatase activity is only relevant after a protein tyrosine kinase phosphorylates the downstream substrate target(s), a screen was undertaken to find neural tyrosine kinases that genetically interact with *Drosophila* LAR. *Abl* mutants emerged as strong, dose-sensitive suppressors of *LAR* ISNb guidance phenotypes.[30] Consistent with these genetic interactions, elevation of Abl activity in a wild type background induces ISNb bypass phenotypes identical to LAR loss.[30] This Abl gain-of-function phenotype is strictly kinase dependent. Thus, increased Abl activity mimics the loss of *Drosophila* LAR in motor axon targeting to the muscle. Further, both Abl and LAR localize to embryonic axons in wild type embryos,[35,36] suggesting direct access in vivo. Although the genetic interactions need not reflect a direct molecular interaction, binding studies reveal that *Drosophila* Abl can associate directly with the

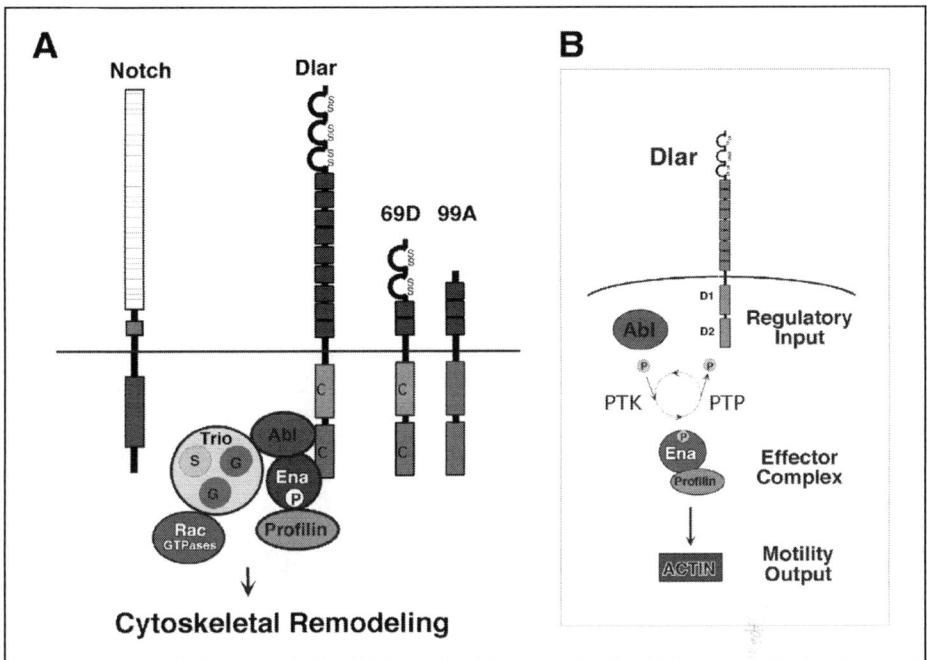

Figure 3. Drosophila signaling machinery required for ventral motor axon guidance. A) A cartoon summary of the Abl-interacting proteins and receptors that control ISNb motor axon guidance in the ventral region of the Drosophila embryo. Direct contact indicates a demonstrated physical interaction, which is also supported by genetic interactions in most cases (see text). All of the intracellular proteins here either bind to actin or are known to regulate actin cytoskeletal assembly or remodeling in some way. B) In the LAR pathway, Abl acts as an antagonist of LAR and Ena, upstream of actin.

LAR cytoplasmic domain (Fig. 3).[30] Indeed, both mammalian Abl and *Drosophila* Abl phosphorylate LAR on the catalytic D2 PTP domain in vitro where Abl has been shown to bind, but not the adjacent D1 domain. Conversely, Abl serves as a good substrate for *Drosophila* LAR PTP activity in vitro.[30] The antagonistic relationship between Abl and LAR is further demonstrated in vivo by the fact that the Abl gain-of-function ISNb bypass phenotype is suppressed by elevating the expression of LAR in postmitotic neurons.[30]

One way that the antagonistic relationship between Abl and LAR in motor axons and their reciprocal catalytic activities could control growth cone guidance is through the regulation of shared phosphoprotein substrates. An attractive candidate is Enabled, the founding member of the Ena/VASP family of actin-binding proteins (Fig. 3).[14,17,37] Ena phosphorylation in vivo is dependent on Abl and the two proteins interact directly through an SH3 domain-binding proline-rich sequence in the central region of Ena.[17,38] Ena also localizes to embryonic axons in wild type, and displays gross defects in the CNS neuropil when disrupted.[17] Moreover, Ena protein binds directly to the D2 PTP domain of LAR and serves as a substrate for LAR in vitro.[30] Indeed, inspection of motor axon pathways in strong zygotic *ena* mutants reveals ISNb motor axon bypass phenotypes highly reminiscent of LAR mutants.[30] Interestingly, Ena is also known to bind directly to Profilin through the same central proline-rich region that associates with SH3-domain of Abl.[39] Together, these observations suggest a model where LAR acts to promote Ena activity in opposition to Abl and Profilin activity (Fig. 3).

Elegant dissection of Ena/VASP functional mechanism in vertebrate cells and in vitro assays suggests that Ena acts to promote actin assembly by protecting the growing tips of microfilaments from capping proteins that terminate actin polymerization.[40] Consistent with this model, intermediate doses of the F-actin-capping drug cytochalasin D induce an ISNb axon bypass phenotype very similar to *ena* loss-of-function, suggesting that an increase in capping activity may compete and interfere with Ena function.[33] Mammalian c-Abl can be induced to phosphorylate mammalian enabled (Mena) in the presence of the Abl-interaction protein Abi.[36] However, it is not known whether Abl kinases have the same functional relationship to Ena/VASP proteins during vertebrate axon guidance decisions.

While the activities of LAR, Abl, Ena and Profilin provide genetic and biochemical links between the growth cone surface and the leading edge cytoskeleton, the intracellular machinery linked to Abl is far more complex. For example, LAR genetically interacts with the small GTPase Rac, known to control cell motility behavior and actin assembly at the leading edge. Dominant-negative Rac (Rac^{N17}) induces an ISNb motor axon bypass phenotype highly reminiscent of LAR mutants,[33] whereas activated Rac (RacV12) induces an ISNb stop short phenotype similar to Abl.[33,41] Simultaneous reduction of Rac and LAR activity is synergistic for ISNb guidance.[33] In addition, LAR displays genetic interactions with the guanine nucleotide exchange factor Trio.[13] Despite the presence of both Rac and Rho-specific exchange factor (dbl-homology) domains in Trio, its function is mediated exclusively via the Rac-specific domain in *Drosophila* retinal axon guidance as well as in *C. elegans*.[42,43] Interestingly, Trio was also identified as one of the original enhancers of Abl lethality, which is suppressed by reduced Ena activity during axonal development in the CNS and results in survival of *Drosophila* adults.[19]

To complicate matters further, additional receptors appear to link ISNb motor axon pathfinding behavior to Abl signaling. Although first known for its central role in the cell-cell interactions that determine cell lineage and neuronal cell fate during neurogenesis, the receptor Notch has been more recently implicated in neuronal differentiation and even synaptic function. Genetic interaction assays show that Notch is a dose-sensitive enhancer of Abl.[25] Using a temperature-sensitive Notch allele to disrupt gene function after cell differentiation reveals that reduction of Notch activity leads to an ISNb motor axon bypass phenotype in *Drosophila*.[26] In this context, Notch displays genetic interactions with many of the same intracellular signaling proteins and effectors previously implicated in the LAR pathway (Fig. 3). Interestingly, the Notch ligand Delta also appears to regulate this process, and has been localized in cells associated with tracheal structures anatomically close to the ISNb axons, suggesting that Delta might be instructive for Notch action during motor axon pathfinding in Drosophila.[26] Future studies will have to address the questions of whether the receptor Notch provides directional information to growth cones and if Notch-dependent signaling events are integrated with other pathways during ISNb guidance. In vertebrates, classes of receptors may also be important for Abl function.[44]

Abl's Role in Axon Guidance at the Midline

Another context where Abl is required for accurate axon guidance is in the formation of the ladder-like scaffold of the *Drosophila* central nervous system (CNS). Axons are guided along an anterior-posterior axis, while axon projections cross the midline where specialized midline glial cells act to orchestrate and restrict the crossing of axons so that neural circuits on the two sides of the animal can be coordinated (reviewed by refs. 45,46). At the early stage of axon guidance in *Drosophila* embryonic CNS, contralateral axons are attracted to the midline glia by secreted Netrins,[47,48] analogous to midline attraction in *C. elegans* and vertebrate embryos (reviewed by ref. 45). Once the CNS axons reach the midline, they continue across the midline to the opposite side. At this point in the guidance process, the ipsilateral growth cones are prevented from inappropriate midline crossing or returning back across the midline by a second secreted factor, Slit,[49] (reviewed by ref. 50). Both Netrins and Slit are expressed by the midline glia, but they act through distinct receptors (Fig. 4). Netrin's attractive response acts via the UNC-40/DCC (Deleted in Colon Cancer) receptor, and Slit repulsion acts via three members of the Roundabout receptor family (Robo, Robo2 and Robo3). Slit activity is also dependent on cell

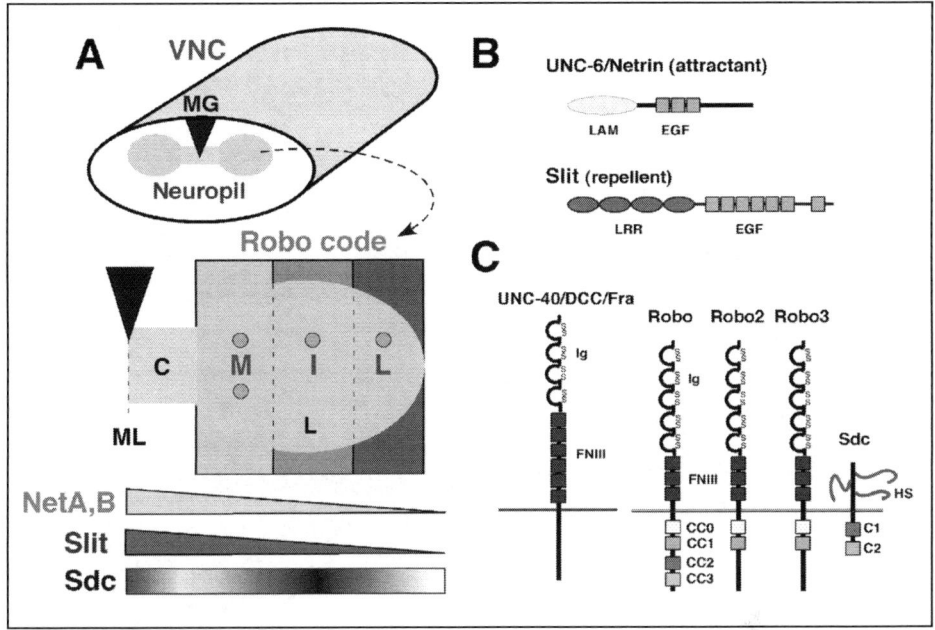

Figure 4. Axonal repellent signaling at the CNS midline. A) The neuropil of the insect CNS (C = commissural axon bundles; L = longitudinal axonal connectives) is shown relative to the position of midline glia (MG) that express the attractive Netrins A and B (gradient emanating from the midline source), as well as the repellent Slit (gradient from midline source). Longitudinal axon surfaces express a combination of Roundabout (Robo) receptors at late stages of embryonic development in addition the UNC-40/DCC-Frazzled receptor. (Robo Code: light grey = Robo alone; medium grey = Robo + Robo3; dark grey = Robo + Robo2 + Robo3) Longitudinal axons also express the proteoglycan Syndecan (Sdc), which is required for efficient repellent responses. B,C) The known secreted midline attractant (UNC-6/Netrin) and repellent (Slit) are diagrammed with domains (LAM = laminin homology domain; EGF = epidermal growth factor-like domain; LRR = leucine-rich repeat domain). Receptors that mediate binding to Netrin and/or Slit are shown (Ig = immunoglobulin domain; FNIII = fibronectin type-III-like domain; CC0-3 = Robo conserved motifs; HS = heparan sulfate polymer; C1-2 = Sdc conserved motifs).

surface proteoglycans like Syndecan that appear to shape the distribution of the guidance factor and may influence ligand-receptor interactions.[51,52] The early repulsive cue that prevents CNS axons from recrossing the midline are mediated primarily by Robo and Robo2 (Fig. 4).[53,54] After the decision to cross or not cross the midline is complete, all three Robo family receptors act in a combinatorial code to determine the lateral position of axons parallel to the midline (the "Robo Code";[53,54] reviewed by ref. 55). The specificity of downstream responses to Netrins and Slit is determined by the cytoplasmic domains of their corresponding receptors, suggesting that the repulsion and attractive guidance apparatus each has a distinct signaling output within the growth cone.[56]

Abl and the Slit Repulsion Pathway

The first indication that Abl is involved in midline guidance came from studies that revealed an antagonistic relationship between Abl and the Slit receptor Robo.[27] Abl overexpression was discovered to mimic loss of midline repulsion via Robo receptor function, resulting in inappropriate midline crossing (Fig. 5). If Abl function was antagonistic to Robo repulsion in the formation of CNS scaffold, then increasing the amount of Abl in a background of reduced Robo function should lead to enhanced defects or more midline crossing.

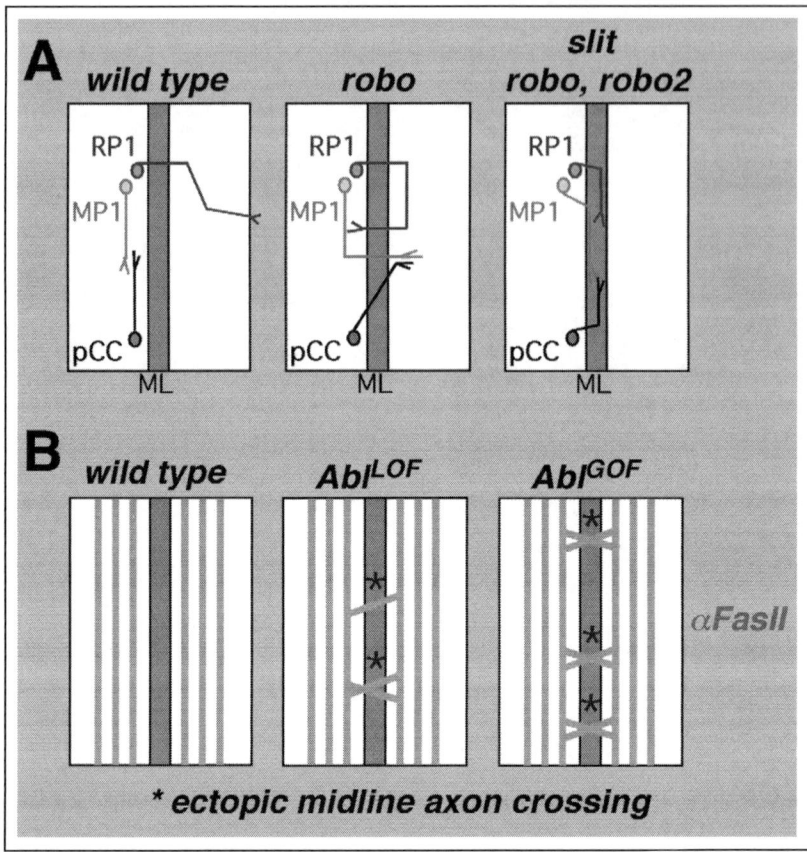

Figure 5. The phenotype of Abl mutants at the CNS midline. A) Identified neurons (RP1, MP1 and pCC) are diagrammed in a top-down fillet view of a *Drosophila* embryonic CNS. When individual axon trajectories are followed, loss of Robo (or Robo2) leads to inappropriate midline axon crossing. In the absence of Slit, or both Robo and Robo2 receptors (*robo,robo2*), all axons collapse into the midline, being overwhelmed by Netrin-based attraction. B) The axon marker Fasciclin II stains multiple bundles of three ipsilateral, longitudinal axons that run parallel on either side of the midline glia. In mutant embryos lacking Abl (LOF) or expressing elevated Abl (GOF) in postmitotic neurons, Fasciclin II-positive axons cross the midline. This phenotype is less severe but qualitatively similar to loss of Robo.

Indeed, this is exactly what was observed with the overexpression of wild type Abl and the modulation of Robo.[27] In addition, Abl was shown to bind to a conserved proline-rich motif (CC3) in the receptor cytoplasmic domain (see Fig. 4 for Robo structure). Mass spectrometry assays also revealed a second conserved Robo motif (CC0) that is phosphorylated by Abl. Interestingly, a phenylalanine substitution at this tyrosine residue renders Robo dominantly active,[27] suggesting that CC0 is a possible site of Abl antagonism. Although CC0 and CC1 are conserved in Robo2 and Robo3, CC3 is not found in these other Robo-family receptors, predicting that the Abl interaction might be specific to Robo.

In a parallel study, Abl zygotic null alleles were shown to mimic loss of midline repulsion, which meant that a reduction in Abl function also caused inappropriate midline crossing;[28] (Fig. 5). These observations suggest that in addition to the antagonism of Robo, Abl is also required to promote Robo signaling. In this study, various double-mutant assays confirmed that Abl acts cooperatively with Slit and all three *Drosophila* Robo receptors.[28] Based on these findings, Abl must have some means of interaction with Robo independent of the CC3 motif, since the CC3 motif is only found in Robo, not in Robo2 or Robo3. These observations have been confirmed by a third independent study on the relationship between the Robo receptors and Abl, where reduction and elevation of Abl in backgrounds with defective repulsion resulted in enhanced midline crossing.[29] In all studies, Abl kinase activity was necessary. Taken together, Abl must play a dual role in the transduction of the Slit repellent pathway.

How does Abl antagonize and promote the Slit Repulsion pathway in midline axon guidance? Studies in the attraction of CNS axons to the midline reveal that activation of Robo by Slit results in silencing of the Netrin receptor UNC-40/DCC/*Drosophila* Frazzled.[57] Frazzled was shown to have direct interaction with both Abl and Trio, a RhoGEF that has two GEF domains that possibly acts on Rac and Rho.[58,59] An extensive network of genetic assays indicate perturbations in Trio and Abl have a negative impact on Frazzled function in vivo, suggesting a model where Abl pathway signaling is also required to facilitate growth cone attraction towards the source of Netrin.[58] These observations reinforce the notion that the Abl kinase pathway is part of core machinery required for both attractive and repellent guidance decisions. However, the study also raises an important question of how to best determine the specificity of Abl genetic interactions in a system like the midline where both types of receptors are thought to engage simultaneously.

Since Abl requires an intact catalytic domain to antagonize and promote the Slit repellent pathway, the phosphorylation of Abl substrates appears to be critical. Robo itself may be the target for an inhibitory feedback mechanism requiring Abl tyrosine phosphorylation. However, other possible downstream substrates required for Robo repulsion are likely to be involved. One key candidate is the Abl substrate Ena.

The Cytoskeletal Connection at the Midline: Actin Effectors

Ena is a member of a conserved family of proteins with an N-terminal EVH1 domain, a central proline-rich region, which binds the actin monomer-binding protein Profilin and several SH3 proteins including Abl, and a C-terminal EVH2 domain that binds F-actin.[37] The Robo CC2 motif is a consensus-binding site for the EVH1 domain of Ena. In vivo and in vitro assays showed that Ena and Robo bind through the CC2 motif, although the CC1 motif is also required for a stable association.[27] In Slit repulsion studies, genetic interactions between Ena and Slit/Robo repulsion components were investigated.[27,28,59] The *Robo;Slit* double mutants display inappropriate midline crossing, as expected since one of the key repellent receptors and its ligand have reduced activity.[27,59] When Ena is reduced in the same *Robo;Slit* background, a dramatic enhancement in midline crossing was observed.[27] This result is consistent with Ena playing a role in midline repulsion, which appears to be a conserved function in *C. elegans* Ena/Unc-34 (Fig. 6).[59]

The Ena family proteins promote actin polymer assembly (reviewed by ref. 60), however, actin polymerization is not synonymous with forward movement of the cell. Indeed, loss of mammalian Ena/VASP-family members increases the velocity of nonneuronal cell migration.[40] Exactly how increased polymerization contributes to growth cone repulsion is currently unknown, since Ena/VASP family proteins have been shown to promote filopodial extension. Since Ena was originally identified as a suppressor of Abl function,[16] it follows that a dose reduction of Ena suppresses the Abl mutant repulsion phenotype, resulting in fewer midline crossings.[28] Interestingly, single *Ena/Unc-34* null mutants display very weak midline phenotypes, indicating that Ena-family proteins play a minor or modulation role in Robo-mediated repulsion.[59] Based on these observations, it is not surprising that several additional actin-associated proteins have been implicated in the midline guidance pathways.[28,61]

Profilin (*Drosophila* chickadee) and cyclase-associated protein (*Drosophila* capulet) have been linked to the Robo signaling pathway by genetic and/or biochemical analyses.[28,62] Although Profilin and the cyclase-associated protein (Capt) can act as actin monomer sequestering factors in some assays, other data suggests a role as an actin monomer delivery proteins that

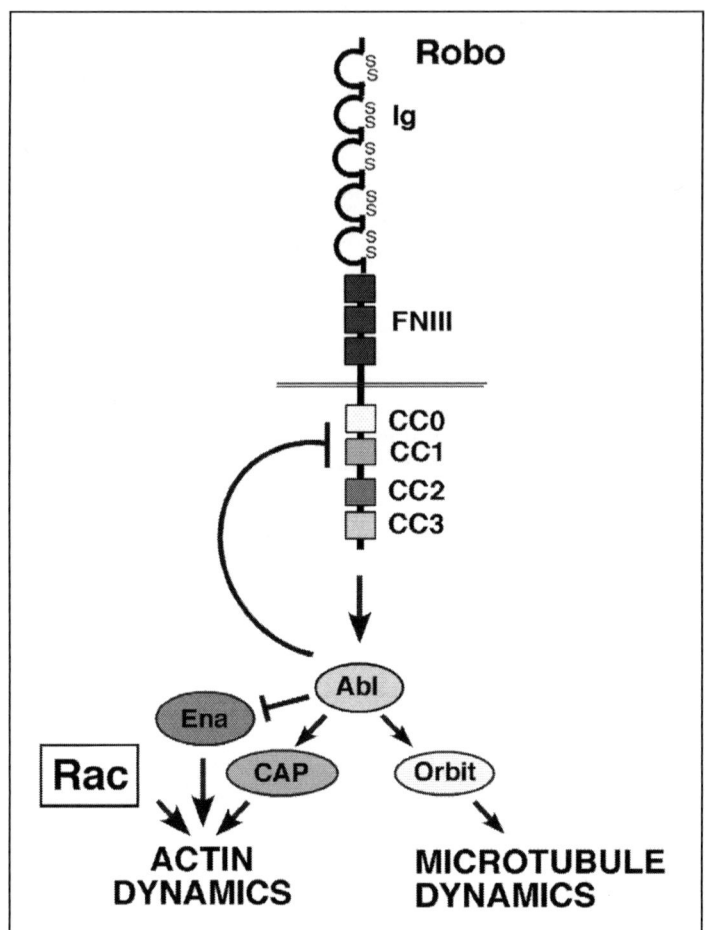

Figure 6. Abl's role in Robo receptor signaling. A cartoon diagram of the Abl-interacting effector proteins that collaborate in the repellent response to Slit. Localization and allied data from nonneuronal studies indicate that these effector proteins control the assembly and dynamics of either actin or tubulin polymers in the growth cone cytoskeleton. Arrows indicate cooperative interactions, whereas Bars indicate antagonistic interactions. There are additional factors that influence the downstream response that have not been directly linked to Abl (e.g., Rac).

promote actin assembly in collaboration with additional actin-binding proteins such as Cofilin (reviewed by ref. 63). Interestingly, hemizygous *Capt* mutants (one wild type copy of the gene and one mutant allele) do not show any detectable CNS repulsion phenotype. However, when only one copy of Capt is removed with only one copy of the Abl gene in the same mutant (transheterozygote), there is an increase in midline crossing.[28] The enhancement in the *Abl* mutant phenotype by a reduction in Capt activity might indicate a common function in promoting repulsion.[28] When only one dose of Capt was combined with *Robo;Robo2* double mutants, midline crossings almost doubled in number.[28] From these results it appears that Capt might function to sequester actin monomers to promote the Robo repulsion at the midline. This might be accomplished by Capt binding to Abl for regulation via the Abl SH3 domain, although this interaction has only been demonstrated in vitro assays in yeast.[64]

Interestingly, *Profilin* (*Chic*) mutants alone do not exhibit any midline crossing defects.[62] However, one dose of Profilin reduces the amount of midline crossing in partial *Robo* mutants, suggesting that Profilin might possibly promote actin assembly to antagonize Robo repulsion and cause attraction.[62] Ironically, in this same study *Chic;Robo* double mutants drastically increase midline crossing of axons to the point of an almost complete loss of Robo repulsion, suggesting the Profilin might aid in inhibiting actin assembly, similar to Capt.[62] Is it possible that Capt and Profilin serve different roles, with Capt promoting repulsion and Profilin antagonizing repulsion of CNS axons to the midline? Is it possible Profilin serves both functions, depending on the presence or absence of regulators? Recall, Ena binds to actin and promotes actin polymerization, however, Ena promotes repulsion. Since Profilin is known to bind Ena family proteins, there may be a link between Ena and Profilin in the Robo pathway. However, Profilin associates with a large number of actin-regulatory proteins, including members of the Formin family and the actin-related protein (Arp2/3) complex, so it may have a number of distinct activities. Clearly, more work will have to be done to address these questions.

Abl Signaling Turns toward Microtubule Dynamics

While the majority of Abl effectors that have been placed in the Slit repellent pathway have been regulators of actin dynamics, recent work demonstrates that Abl is also coordinating the dynamics of microtubule polymers in the growth cone (Fig. 6). In a recent gain-of-function screen[65] using overexpression of Abl to discover novel Abl effectors in *Drosophila*, the microtubule associated Orbit/MAST (Multiple asters) was identified.[61]

In this screen, the UAS/GAL4 transcriptional activator system[66] was used to elevate Abl and Orbit/MAST expression in the *Drosophila* retina, revealing a synergistic interaction between the two genes(refer to Fig. 2B).[61] The Abl signaling partner Orbit/MAST is homologous to CLASP1 and CLASP2, members of the microtubule plus-end tracking ("+TIP") class of proteins.[67] Loss of zygotic Orbit/MAST leads to ectopic midline axon crossing very similar to loss of Abl. Although Orbit/MAST is important for mitotic spindle formation during cell division, its role in midline axon guidance is strictly postmitotic.[61] At the midline, Orbit/MAST cooperates with Slit, Robo and Robo2, but not Robo3.[61] In addition, Orbit/MAST localizes within *Drosophila* growth cones. Dynamic imaging in Xenopus neurons revealed that CLASP2 selectively associates with a subset of growth cone microtubules. These microtubules are known as pioneers since they penetrate the actin-rich periphery of the leading edge and individual filopodia.[61] CLASP-decorated microtubule plus ends track along actin bundles that support individual filopodia, placing this Abl partner at the interface between pioneer microtubules and the actin cytoskeleton. Interestingly, vertebrate Abl-related gene (Arg) kinase, binds to both actin and microtubule polymers, placing it at the same interface as CLASP.[68]

Other Outputs of the Abl Kinase

Additional experiments in cell culture looked at the role of Robo and Abl in modulating cell adhesion mechanisms. This has implicated Abl in Slit-Robo mediated signaling via ß-Catenin which downregulates cadherin-dependent cell adhesion.[69] Activation of Robo leads to phosphorylation of β-Catenin, an effect that reduces catenin-cadherin interactions and thus induces a decrease in cell adhesion. This phosphorylation of β-Catenin by Robo is blocked by the Abl inhibitor Gleevec (also known as STI571 [Novartis]), suggesting that Abl activation is key to the modulation of cell adhesion by Slit.[69] Further, an interaction has been observed between delta-Catenin and Abl in vertebrate neurons.[70] Interestingly, *Drosophila* β-catenin (Armadillo) isoforms show striking genetic interactions with Abl during axonal patterning in the embryonic CNS.[71] Although the importance of the Robo-Cadherin mechanism in vivo is not known, these data suggest yet another way in which Abl may act to coordinate the many aspects of growth cone cell biology involved in directional axon guidance (reviewed by ref. 72). These observations reinforce the notion that the Abl kinase pathway is part of core machinery required for both attractive and repellent guidance decisions.

Abl's Role in Dendrites and Synapses

Abl and Arg Modulate Dendritic Morphogenesis

Abl function is not only necessary for the axon to negotiate the complex terrain of multiple extracellular cues, this transducer has been shown to play an important role in establishing synaptic connection with its target postsynaptic membrane. Dendrites are dynamic, specialized postsynaptic processes where neurons receive, process, and integrate signals from their presynaptic partners for branching, retraction, elimination, and maintenance.[73] The morphology of a population of dendrites can range from simple to intricate branching depending on both extrinsic and intrinsic signals unique for its neural environment and function (e.g., see ref. 74). Although equally important for the formation of functional neural circuits, far less is known about the formation of dendrites compared to our understanding of axonal development. While *Drosophila* is rapidly proving to be an attractive model system for work in dendrite morphogenesis, our current understanding of Abl's contribution in this process comes primarily from work in mammals. Recent genetic experiments from Koleske and coworkers demonstrate that murine Abl and Arg act semi-redundantly in cortical neurons to regulate dendritic branching in vitro and in vivo.[75]

Using conditional knock-out technology to circumvent the requirement for the two kinases for neurulation (e.g., see ref. 2), it was possible to examine cortical development in *Abl,Arg* double mutant mice. These juvenile *Abl,Arg* double mutant animals display a significant decrease in the number and branch complexity of both apical and basal dendrites. Interestingly, the defect corresponds to a late onset effect on the maintenance of dendritic projections, as young animals (~P21) have normal dendritic arbors (Fig. 7).[75] In vitro studies suggest that the major role of Arg is to promote dynamic growth of dendrites from cortical neurons, an effect that is revealed on a laminin-rich substrate, suggesting a link between Arg and integrin-mediated substrate adhesion.[75] Indeed, Abl and Abl-interactor (Abi) were shown to colocalize with catenin in neural cultures, while Arg phosphorylation of the Rho activator, p190RhoGAP required integrin and cadherin cell adhesion.[70,76,77] Taken together, these results suggest that the cytoskeletal functions of Abl and Arg are coordinated with an extracellular matrix to drive dendritic remodeling via actin regulators (Fig. 7). This allows the progression and maturation of cortical circuitry and provides a mechanism to link cell-substrate adhesion and Abl activation to the cytoskeletal remodeling likely to underlie dendritic growth.

The in vivo experiments of Koleske and colleagues confirm earlier data from hippocampal cell culture, in which the Abl inhibitor STI571 was used to implicate Abl in dendritic development.[78] When STI571 was applied to cultures prior to the establishment of dendritic identity, a noticeable reduction in dendrite length and secondary branching was observed.[78] Conversely, expression of constitutively active Abl results in a general increase in dendritic complexity.[78] Indeed, in this same study coexpression of constitutively active Abl and RhoA resulted in the RhoA phenotype of dendrite retraction and simplification of dendritic complexity, suggesting RhoA functions downstream of Abl.[78] Moreover, Arg was recently shown to stimulate p190RhoGAP, which is an upstream regulator of RhoA, resulting in the inhibition of neurite outgrowths in neural cultures.[76] This occurs by the Arg phosphorylation of p190RhoGAP.[75,88] Thus, Arg may limit the postsynaptic dendritic development by signaling through a RhoA pathway.[75]

The findings of Jones and colleagues are also consistent with analysis of the ectopic expression of dominant-negative and activated small Rho GTPases.[73,74,77] In these studies, the expression of activated RhoA in cortical or hippocampal slice cultures caused a significant reduction in dendritic branching, while perturbation of RhoA by C3 transferase or p190RhoGAP increased branching. When endogenous RhoA activity was abrogated later in development, dendritic branching remained unchanged, remarkably similar to the late onset defect observed in *Abl,Arg* double mutants by Koleske and his colleagues.[73,75]

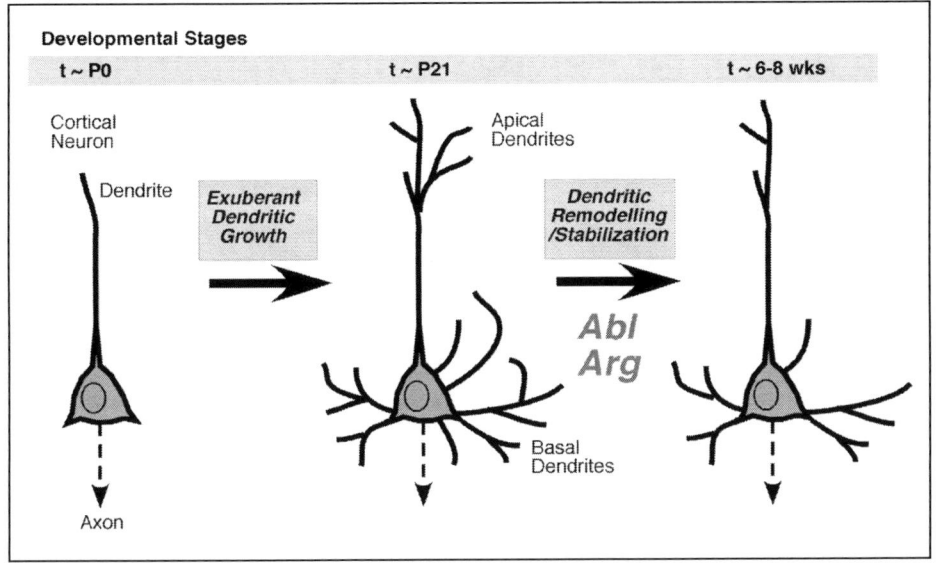

Figure 7. Abl regulates dendritic branching. A temporal scheme is shown for the elaboration and remodeling of dendritic branches from cortical neurons in the mammalian brain (time line above diagrams). Postnatal development of apical and basal dendrites and dendritic branches occurs from P0 to approximately P21, prior to the remodeling of dendritic arbors in the weeks that follow infancy. Abl and Arg appear to control the dynamics of dendritic remodeling and are necessary for the persistence of dendrites at later stages of brain development.

The action of Rho GTPases are likely to be conserved across species. Looking at RhoA function in the mushroom body of the *Drosophila* brain, defects have been observed in the stereotypical pattern of axonal and dendritic morphology.[79] Expression of constitutively active RhoA in *Drosophila* resulted in a reduction in the dendritic volume, while RhoA mutants showed overextension of their dendrites outside the normal growth region.[79] Although this *Drosophila* study does not examine Abl function directly, these results confirm the role of RhoA in limiting dendritic outgrowth or activating retraction.[73] This data strongly indicates that a conserved Abl and/or Arg function is necessary in dendrite development through the RhoA signaling pathway. In another *Drosophila* study a mutant screen was used to identify multiple proteins involved in the formation of dendrites in the peripheral nervous system.[80] One of the molecules identified was Ena. Interestingly, *Drosophila Ena* larval mutants had normal primary dorsal dendritic branches, however, the secondary lateral branches were overextended and misdirected, turning dorsally.[80]

Abl Kinases at the Synapse

The previous sections have presented evidence of Abl and Arg function during neural development. Abl and Arg are expressed in the brain, with immunoelectron and light microscopy localizing both to the murine hippocampus.[81] In particular, Abl and Arg expression were detected in presynaptic densities and at dendritic postsynaptic structures.[2,81,91] Although Arg null mice show no gross defects in neural development and live to adulthood, the importance of both Abl and Arg function is evident in *Abl/Arg* double knockout mice where the neural tube collapse occurs, resulting in embryonic death.[2]

At the protein level, Abl and Arg accumulate on the pre and post-synaptic sides of excitatory glutamatergic dendritic spine synapses in the murine hippocampal CA1 area (Fig. 8).[81] Interestingly, short-term synaptic plasticity is defective in Abl and in Arg single knockout mice

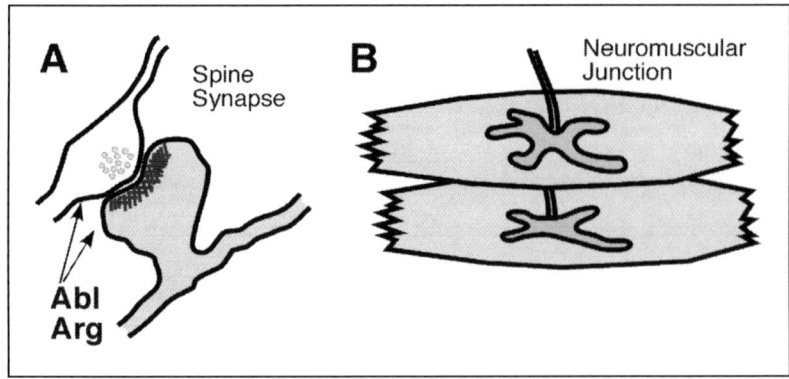

Figure 8. Abl regulates the form and function of synapses. A) Abl and Arg localize to vertebrate central excitatory synapses where glutamatergic terminals (white with synaptic vesicles) contact dendritic spines (grey with postsynaptic density). B) Abl kinase activity is also required at the vertebrate neuromuscular junction. A cholinergic synapse forms between motor axons and muscle fibers where Acetylcholine receptors (AChRs) cluster on the muscle surface at points of nerve contact.

without alterations in basal synaptic transmission.[81] Long-term potentiation (LTP), long-term depression (LTD) and post-tetanic potentiation (PTP) were normal in the single kinase mutants.[81] These results were confirmed by the bath application of the Abl/Arg inhibitor STI571 to wild-type hippocampal slice cultures.[81] However, in the single null mice, paired-pulse facilitation (PPF) was significantly reduced, suggesting that Abl and Arg are both necessary. PPF is a measure of the enhanced response characteristic of the second stimulus in a rapid pair, due to residual calcium in the presynaptic terminal after the first stimulus.[81] PPF was also reduced when wild type hippocampal slice cultures were incubated in STI571, indicating that catalytic activity is required for this aspect of physiological regulation.[81] Based on these results, it appears that Abl and Arg participate in redundant pathways that aid in the release of synaptic vesicles.[81] Although a role for mammalian Abl kinases in the morphogenesis of spines and in CNS synapse assembly has not been reported, it is intriguing that loss of the Abl-interacting protein Abi has a dramatic effect on dendritic spine morphology and LTP in the murine hippocampus.[77]

To date, the best-characterized role for Abl-family kinases during synaptogenesis is at the vertebrate neuromuscular junction (NMJ; reviewed by ref. 82). At these large cholinergic synapses, Abl is important for the postsynaptic clustering of acetylcholine receptors (AchR) triggered by the secreted factor Agrin and the downstream receptor tyrosine kinase MuSK.[83] Both Abl and Arg are expressed in murine muscle.[83] When murine myoblast cells were treated with STI571, the induction of clustering by agrin was inhibited, suggesting the requirement of Abl function.[83] In vivo experiments using a kinase-inactive Abl acetylcholine receptors also failed to cluster.[83] Further biochemical studies demonstrated direct interactions between Abl and the MuSK receptor in the formation of a stable complex.[83] Indeed, mice lacking Arg exhibit muscle difficulties similar to human conditions such as congenital myasthenia. These conditions are thought to be due to defective synaptic transmission defects, possibly indicative of the electrophysiological defects observed in Arg and Abl single mutant mice.[81,83]

Relevance to Human Neurological Development

Abl and Arg are important transducers of developmental signals that require cytoskeletal remodeling as a key part of their output. In recent years, it has become apparent that Abl kinases serve as sites where multiple inputs converge, are processed, directed to the appropriate signal output, and, ultimately, result in a dynamic change in cellular shape. This is particularly true in the development of the neuron. Each developmental process described in the previous sections requires cytoskeletal remodeling of the cell: cell polarity, where neurons identify axons and dendrites, the growth and guidance of axons and dendrites to their targets, where axons and dendrites must orchestrate a precisely apposition with its synaptic surface. Moreover, recent evidence has revealed how efficient actin machinery is crucial to human cognitive development. Indeed, many causes of mental retardation stem from mutations in key effector molecules resulting in impairment in memory, learning, socialization and general intelligence. Although Abl is only known for its role in the human leukemias, there are several forms of X-linked nonsyndromic mental retardation (MRX) that have been mapped to genes involved in signal pathways that regulate actin dynamics.[59,84-86] This form of mental retardation is characterized by grossly normal brain development with little or no obvious defects.[84]

There are possibly as many as 65 genes on the X chromosome that cause nonsyndromic mental retardation, most of which have not been characterized.[59] One of the genes is oligophrenin, a protein that was shown to have Rho GTPase GAP activity for Rac, Cdc42 and Rho.[84] The Rho GTPases have been implicated in axon guidance and neurite outgrowth.[33,73-77,79] More specifically, Arg was recently shown to activate the RhoA inhibitor, p190RhoGAP.[76] PAK3 is another MRX gene mapped to a locus on the X chromosome.[59,84] PAK3 is a serine/threonine kinase that is activated by both Rac and Cdc42 RhoGTPases, and whose activity is essential for cortical development.[70] In addition, the p35-Cdk5 complex, that can bind PAK and inhibit it function, is also activated by Rac.[70] Interestingly, Abl forms a complex with Cable and Cdk5 for function.[87] A third MRX gene, ARHGEF6, encodes a guanine nucleotide exchange factor or GEF for Rho GTPases.[59] Hence, either defective increases or decreases in RhoGTPase signaling results in cognitive limitations.[59] Further, mutations in a Rho GAP called MEGAP that is related to the Robo-receptor signaling effector srGAP have been linked to mental retardation and ataxia in human patients.[88] Moreover, mutations in human Robo3/Rig have been linked to human horizontal gaze palsy with progressive scoliosis (HGPPS) involving defects in axon pathway formation.[89,90] While there is not yet direct evidence that Abl plays a role in any of these genetic conditions, the known relationships between Abl and these other molecules makes Abl a likely candidate for future studies in this area of study.

Acknowledgements

D. Van Vactor is a Leukemia and Lymphoma Society Scholar, and is supported by grants from the National Institute of Neurological Disease and Stroke. C.L. Thompson is supported by a training grant from the National Cancer Institute (T32CA09361).

References

1. Lee H, Van Vactor D. Neurons take shape. Curr Biol 2003; 13:R152-161.
2. Koleske AJ, Gifford AM, Scott ML et al. Essential roles for the Abl and Arg tyrosine kinases in neurulation. Neuron 1998; 21:1259-1272.
3. Hoffman-Falk H, Einat P, Shilo BZ et al. Drosophila melanogaster DNA clones homologous to vertebrate oncogenes: Evidence for a common ancestor to the src and abl cellular genes. Cell 1983; 32:589-598.
4. Hoffmann FM, Fresco LD, Hoffman-Falk H et al. Nucleotide sequences of the Drosophila src and abl homologs: Conservation and variability in the src family oncogenes. Cell 1983; 35:393-401.
5. Henkemeyer MJ, Gertler FB, Goodman W et al. The drosophila ableson proto-oncogene homologue: Identification of mutant alleles that have pleiotropic effects late in development. Cell 1987; 51:821-828.
6. Grevengoed EE, Loureiro JJ, Jesse TL et al. Abelson kinase regulates epithelial morphogenesis in Drosophila. J Cell Biol 2001; 155:1185-1198.
7. Hoffmann FM. Drosophila abl and genetic redundancy in signal transduction. Trends Genet 1991; 7:351-355.
8. Wills Z, Marr L, Zinn K et al. Profilin and the Abl tyrosine kinase are required for motor axon outgrowth in the drosophila embryo. Neuron 1999a; 22:291-299.
9. Keshishian H, Broadie K, Chiba A et al. The drosophila neuromuscular junction: A model system for studying synaptic development and function. Annu Rev Neuroscience 1996; 19:545-575.

10. Van Vactor D, Sink H, Fambrough D et al. Genes that control neuromuscular specificity in Drosophila Cell 1993 73:1137-1153.
11. Broadie KS, Bate M. Development of the embryonic neuromuscular synapse of Drosophila melanogaster. J Neurosci 1993; 13:144-66.
12. Lee S, Harris KL, Whitington PM et al. Short stop is allelic to kakapo, and encodes rod-like cytoskeletal-associated proteins required for axon extension. J Neurosci 2000; 20:1096-108.
13. Bateman J, Van Vactor D. The Trio family of guanine-nucleotide-exchange factors: Regulators of axon guidance. J Cell Sci 2001; 114(Pt 11):1973-80.
14. Gertler FB, Bennett RL, Clark MJ et al. Drosophila abl tyrosine kinase in embryonic CNS axons: A role in axonogenesis is revealed through dosage-sensitive interactions with disabled. Cell 1989; 58:103-113.
15. Gertler FB, Hill KK, Clark MJ et al. Dosage-sensitive modifiers of drosophila abl tyrosine kinase function: Prospero, a regulator of axonal outgrowth, and disabled, a novel tyrosine kinase Substrate. Genes Dev 1993; 7:441-453.
16. Gertler FB, Doctor JS, Hoffman FM. Genetic suppression of mutations in the drosophila abl proto-oncogene homologue. Science 1990; 248:857-248.
17. Gertler FB, Comer AR, Juang J-L et al. Enabled, a dosage-sensitive suppressor of mutations in the drosophila Abl tyrosine kinase, encodes an Abl substrate with SH3 domain-binding properties. Genes Dev 1995; 9:521-533.
18. Hill KK, Bedian V, Juang J-L et al. Genetic interactions between the drosophila ableson (Abl) tyrosine kinase and failed axon connections (Fax), a novel protein in axon bundles. Genetics 1995; 141:595-606.
19. Liebl EC, Forsthoefel DJ, Franco LS et al. Dosage-Sensitive, reciprocal genetic interactions between the abl tyrosine kinase and the putative GEF trio reveal trio's role in axon pathfinding. Neuron 2000; 26:107-118.
20. Liebl EC, Rowe RG, Forsthoefel DJ et al. Interactions between the secreted protein Amalgam, its transmembrane receptor Neurotactin and the Abelson tyrosine kinase affect axon pathfinding. Development 2003; 130:3217-3226.
21. Woodring PJ, Litwack ED, O'Leary DD et al. Modulation of the F-actin cytoskeleton by c-Abl tyrosine kinase in cell spreading and neurite extension. J Cell Biol 2002; 156:879-892.
22. Elkins T, Zinn K, McAllister L et al. Genetic analysis of a drosophila neural cell adhesion molecule: Interaction of fasciclin I and abelson tyrosine kinase mutations. Cell 1990; 60:565-575.
23. Garcia-Alonso L, VanBerkum MF, Grenningloh G et al. Fasciclin II controls proneural gene expression in Drosophila. Proc Natl Acad Sci USA 1995; 92:10501-10505.
24. Hu S, Sonnenfeld M, Stahl S et al. Midline Fasciclin: A Drosophila Fasciclin-I-related membrane protein localized to the CNS midline cells and trachea. J Neurobiol 1998; 35:77-93.
25. Giniger E. A role for Abl in Notch signaling. Neuron 1998; 20:667-681.
26. Crowner D, Le Gall M, Gates MA et al. Notch steers drosophila ISNb motor axons by regulating the Abl signaling pathway. Curr Biol 2003; 13:967-972.
27. Bashaw GJ, Kidd T, Murray D et al. Repulsive axon guidance: Abelson and enabled play opposing roles downstream of the roundabout receptor. Cell 2000; 101:703-715.
28. Wills Z, Emerson M, Rusch J et al A Drosophila homologue of cyclase-Associated-Proteins collaborates with the Abl tyrosine kinase to control midline axon pathfinding. Neuron 2002; 36:611-622.
29. Hsouna A, Kim YS, VanBerkum MF. Abelson tyrosine kinase is required to transduce midline repulsive cues. J Neurobiol 2003; 57:15-30.
30. Wills Z, Bateman J, Korey C et al. The tyrosine kinase Abl and its substrate enabled collaborate with the receptor phosphatase dlar to control motor axon guidance. Neuron 1999b; 22:301-312.
31. Streuli M, Krueger NX, Tsai AY et al. A family of receptor-linked protein tyrosine phosphatases in humans and Drosophila. Proc Natl Acad Sci USA 1989; 86:8698-702.
32. Krueger NX, Van Vactor D, Wan HI et al. The transmembrane tyrosine phosphatase DLAR controls motor axon guidance in Drosophila. Cell 1996; 84:611-22.
33. Kaufmann N, Wills ZP, Van Vactor D. Drosophila rac1 controls motor axon sguidance. Development 1998; 125:453-461.
34. Desai CJ, Krueger NX, Saito H et al. Competition and cooperation among receptor tyrosine phosphatases control motoneuron growth cone guidance in Drosophila. Development 1997; 124:1941-52.
35 Henkemeyer M, West SR, Gertler FB et al. A novel tyrosine kinase-independent function of drosophila abl correlates with proper subcellular localization. Cell 1990; 63:949-960.
36. Tani K, Sato S, Sukezane T et al. Abl interactor 1 promotes tyrosine 296 phosphorylation of mammalian enabled (Mena) by c-Abl kinase. J Biol Chem 2003; 278:21685-21692.

37. Gertler FB, Niebuhr K, Reinhard M et al. Mena, a relative of VASP and drosophila enabled, is implicated in the control of microfilament dynamics. Cell 1996; 87:227-239.
38. Comer A R, Ahern-Djamali SM, Juang J-L et al. Phosphorylation of enabled by the drosophila abelson tyronsine kinase regulates the In vivo function and protein-protein interactions of enabled. Mol Cell Biol 1998; 18:152-160.
39. Ahern-Djamali SM, Bachmann C, Hua P et al. Identification of profilin and src homology 3 domains as binding partners for Drosophila enabled. Proc Natl Acad Sci USA 1999; 96:4977-4982.
40. Bear JE, Svitkina TM, Krause M et al. Antagonism between Ena/VASP proteins and actin filament capping regulates fibroblast motility. Cell 2002; 109:509-21.
41. Sone M, Hoshino M, Suzuki E et al. Still life, a protein in synaptic terminals of drosophila homologous to GDP-GTP exchangers. Science 1997; 275:543-547.
42. Steven R, Kubiseski TJ, Zheng H et al. UNC-73 activates the Rac GTPase and is required for cell and growth cone migrations in C. elegans. Cell 1998; 92:785-95.
43. Newsome TP, Schmidt S, Dietzl G et al. Trio combines with dock to regulate Pak activity during photoreceptor axon pathfinding in Drosophila Cell 2000; 101:283-94
44. Yu HH, Zisch AH, Dodelet VC et al. Multiple signaling interactions of Abl and Arg kinases with the EphB2 receptor. Oncogene 2002a; 20:3995-4006.
45. Tessier-Lavigne M, Goodman CS. The molecular biology of axon guidance. Science 1996; 274:1123-33.
46. Dickson BJ. Molecular mechanisms of axon guidance. Science 2002; 298:1959-64.
47. Mitchell KJ, Doyle JL, Serafini T et al. Genetic analysis of Netrin genes in Drosophila: Netrins guide CNS commissural axons and peripheral motor axons. Neuron 1996; 17:203-15.
48. Harris R, Sabatelli LM, Seeger MA. Guidance cues at the Drosophila CNS midline: Identification and characterization of two Drosophila Netrin/UNC-6 homologs. Neuron 1996; 17:217-28.
49. Kidd T, Bland KS, Goodman CS. Slit is the midline repellent for the robo receptor in Drosophila. Cell 1999; 96:785-94.
50. Flanagan JG, Van Vactor D. Through the looking glass: axon guidance at the midline choice point. Cell 1998; 92:429-32.
51. Johnson KG, Ghose A, Epstein E et al. Axonal heparan sulfate proteoglycans regulate the distribution and efficiency of the repellent slit during midline axon guidance. Curr Biol 2004; 14:499-504.
52. Steigemann P, Molitor A, Fellert S et al. Heparan sulfate proteoglycan syndecan promotes axonal and myotube guidance by slit/robo signaling. Curr Biol 2004; 14:225-230.
53. Simpson JH, Bland KS, Fetter RD et al. Short-range and long-range guidance by Slit and its Robo receptors: A combinatorial code of Robo receptors controls lateral position. Cell 2000; 103:1019-32.
54. Rajagopalan S, Nicolas E, Vivancos V et al. Links Crossing the midline: Roles and regulation of Robo receptors. Neuron 2000; 28:767-77.
55. Rusch J, Van Vactor D. New Roundabouts send axons into the Fas lane. Neuron 2000; 28: 637-40.
56. Bashaw GJ, Goodman CS. Chimeric axon guidance receptors: The cytoplasmic domains of slit and netrin receptors specify attraction versus repulsion. Cell 1999; 97:917-26.
57. Stein E, Zou Y, Poo M et al. Binding of DCC by netrin-1 to mediate axon guidance independent of adenosine A2B receptor activation. Science 2001; 291:1976-82.
58. Forsthoefel DJ, Liebl EC, Kolodziej PA et al. The Abelson tyrosine kinase, the Trio GEF and Enabled interact with the Netrin receptor Frazzled in Drosophila. Development 2205; 132:1983-94.
59. Yu TW, Hao JC, Lim W et al. Shared receptors in axon guidance: SAX-3/Robo signals via UNC-34/ Enabled and a Netrin-independent UNC-40/DCC function. Nature Neuroscience 2002b; 5(11):1147-54.
60. Lanier LM, Gertler FB. From Abl to actin: The role of the Abl tyrosine kinase and its associated proteins in growth cone motility. Curr Opin Neurobiol 2000; 10:80-87.
61. Lee H, Engel U, Rusch J et al. The microtubule plus end tracking protein Orbit/MAST/CLASP acts downstream of the tyrosine kinase Abl in mediating axon guidance. Neuron 2004; 42:913-926.
62. Kim Y-S, Furman S, Sink H et al. Calmodulin and profilin coregulate axon outgrowth in drosophila J Neurobiol 2001; 47:26-38.
63. Hubberstey AV, Mottillo EP. Cyclase-associated proteins: CAPacity for linking signal transduction and actin polymerization. FASEB J 2002; 16:487-489.
64. Freeman NL, Lila T, Mintzer KA et al. A conserved proline-rich region of the Saccharomyces cerevisiae cyclase-associated protein binds SH3 domains and modulates cytoskeletal localization. Mol Cell Biol 1996; 2:548-556.

65. Rorth P, Szabo K, Bailey A.et al Systematic gain-of-function genetics in Drosophila. Development 1998; 125:1049-57.
66. Brand AH, Perrimon N. Targeted gene expression as a means of altering cell fates and generating dominant phenotypes. Development 1993; 118:401-415.
67. Akhmanova A, Hoogenraad CC, Drabek K et al. Clasps are CLIP-115 and -170 associating proteins involved in the regional regulation of microtubule dynamics in motile fibroblasts. Cell 2001; 104:923-35.
68. Miller AL, Wang Y, Mooseker MS et al. The Abl-related gene (Arg) requires its F-actin: Microtubule crosslinking activity to regulate lamallipodial dynamics during firboblast adhesion. J Cell Biol 2004; 165(3):407-419.
69. Rhee J, Mahfooz NS, Arregui C et al. Activation of the repulsive receptor Roundabout inhibits N-cadherin-mediated cell adhesion. Nat Cell Biol 2002; 4:798-805.
70. Lu Q, Mukhopadhyay NK, Griffin JD et al. Brain armadillo protein delta-catenin interacts with Abl tyrosine kinase and modulates cellular morphogenesis in response to growth factors. J Neurosci Res 2002; 67:618-624.
71. Loureiro J, Peifer M. Roles of Armadillo, a Drosophila catenin, during central nervous system development. Curr Biol 1998; 8:622-632.
72. Emerson MM, Van Vactor D. Robo is Abl to block N-Cadherin function. Nat Cell Biol 2002; 4:E227-E230.
73. Nakayama AY, Harms MB, Luo L. Small GTPases rac and rho in the maintenance of dendritic spines and branches in hippocampal pyramidal neurons. J Neurosci 2000; 20(14):5329-5338.
74. Threadgill R, Bobb K, Ghosh A. Regulation of dendritic growth and remodeling by rho, rac, and Cdc42. Neuron 1997; 19:625-634.
75. Moresco EM, Donadlson S, Williamson A et al. Integrin-mediated dendrite branch maintenance requires abelson (Abl) family kinases. J Neurosci 2005; 25:6105-6118.
76. Hernandez SE, Settleman J, Koleske AJ. Adhesion-dependent regulation of p190RhoGAP in the developing brain by the Abl-related gene tyrosine kinase. Curr Biol 2004; 14:691-696.
77. Grove M, Demyanenko G, Echarri A et al. ABI2-deficient mice exhibit defective cell migration, aberrant dendritic spine morphogenesis, and deficits in learning and memory. Mol Cell Biol 2005; 24:10905-10922.
78. Jones SB, Lu HY, Lu Q. Abl tyrosine kinase promotes dendrogenesis by inducing actin cytoskeletal rearrangements in cooperation with rho family small GTPases in hippocampal neurons. J Neurosci 2004; 24:8510-8521.
79. Lee T, Winter C, Marticke SS et al. Essential roles of drosophila RhoA in the regulation of neuroblast proliferation and dendritic but Not axonal morphogenesis. Neuron 2000; 25:307-316.
80. Gao R-B, Brenman JE, Jan LY et al. Genes regulating dendritic outgrowth, branching, and routing in Drosophila. Genes Dev 1999; 13:2549-2561.
81. Moresco EM, Scheetz AJ, Bornmann WG et al. Abl family nonreceptor tyrosine kinases modulate short-term synaptic plasticity. J Neurophysiol 2003; 89:1678-1687, (82).
82. Burden SJ, Fuhrer C, Hubbard SR. Agrin/MuSK signaling: Willing and Abl. Nat Neurosci 2002; 6:653-654.
83. Finn AJ, Feng G, Pendergast AM. Postsynaptic requirement for Abl kinases in assembly of the neuromuscular junction. Nat Neurosci 2003; 6:717-723.
84. Allen KM, Gleeson JG, Bagrodia S et al. PAK3 mutation in nonsyndromic X-linked mental retardation. Nat Genet 1998; 20:25-30.
85. Billuart P, Bienvenu T, Ronce N et al. Oligophrenin-1 encodes a rhoGAP protein involved in X-linked mental retardation. Nature 1998; 392:923-926.
86. Chelly J. Breakthroughs in molecular and cellular mechanisms underlying X-linked mental retardation. Human Mol Genet 1999; 8(10):1833-1838.
87. Zukerberg LR, Patrick GN, Nikolic M et al. Cables Links Cdk5 and c-Abl and Facilitates Cdk5 Tyrosine Phosphorylation, Kinase Upregulation, and Neurite Outgrowth. Neuron 2000; 26:633-646.
88. Endris V, Wogatzky B, Leimer U et al. The novel Rho-GTPase activating gene MEGAP/srGAP3 has a putative role in severe mental retardation. PNAS 2002; 99(18):11754-11759.
89. Butcher J. Mutations in ROBO3 cause HGPPS. Lancet Neurol 2004; 3(6):328.
90. Jen JC, Chan WM, Bosley TM et al. Mutations in a human ROBO gene disrupt hindbrain axon pathway crossing and morphogenesis. Science 2004; 304:1509-1513
91. Steel M, Moss J, Clark KA et al. Gene-trapping to identify and analyze genes expressed in the mouse hippocampus. Hippocampus 1998; 8:444-457.

Index